HANDBOOK OF LAWN MOWER REPAIR

HANDBOOK OF LAWN MOWER REPAIR

(updated edition)

FRANKLYNN PETERSON

HAWTHORN BOOKS, INC.
A Howard & Wyndham Company
NEW YORK

DEDICATION

For my father, who first planted these seeds of curiosity, and for Owen, who helped to cultivate them.

HANDBOOK OF LAWN MOWER REPAIR
 All inquiries should be addressed to Emerson Books, Inc., Reynolds Lane, Buchanan, New York 10511. This book was manufactured in the United States of America and published simultaneously in Canada by Prentice-Hall of Canada, Ltd., 1870 Birchmount Road, Scarborough, Ontario.

Library of Congress Catalog Card Number: 77–92313
ISBN: 0–8015–3256–6

1 2 3 4 5 6 7 8 9 10

TABLE OF CONTENTS

PREFACE

When this project began, I was amazed that there were no books available, at least none listed in *Books in Print,* on the subject of lawn mowers. There are, after all, 40 million of them in use almost every sunny Saturday afternoon all summer long. There are also almost 125,000 serious accidents involving lawn mowers every year.

Now that this lawn mower project is finished, I can see why no one apparently has tackled lawn mowers as a subject needing investigation for the millions of consumers who use them. With rare and notable exceptions, industry cooperation with this independent study on lawn mower use and repair was extremely limited. Indeed, the industry itself has only made a very minimal attempt at enforcing a weak standard of safety on member firms within its ranks. Even the traditional consumer publications have shied away from comprehensive investigations of lawn mowers, probably because of the subject's complexity.

This book does not attempt to tell all there is to know about every lawn mower ever sold. There can be no definitive book on the subject because the equipment on sale is in a steady state of flux. Rather than attempting to prepare a catalog of every nut and bolt on every one of the 500 or so different lawn mowers available in North America, I have chosen to develop a broad framework which should be of general assistance to a consumer who wants to or must tackle the sundry problems connected with buying, using and above all, repairing a lawn mower.

Hopefully, after digesting the techniques and guidelines offered here, the consumer will be able to bend successfully with the myriad of variations as they appear. If I dare to make the analogy, this is not a cookbook full of ingredient lists for hundreds of cakes and casseroles, but a book on how to cook even when there might not be a recipe close by. Although it is considered bad form in the sales-oriented publishing world, whenever they are available, I have cited inexpensive and easy to obtain official repair manuals for manufacturers of various lawn mowers or engines used to power lawn mowers.

I have included over a hundred specific illustrations and many generalized ones. Dozens of actual brand names of specific lawn mowing equipment are named and discussed, but the emphasis has not been to draw up an exhaustive (and exhausting) rogues gallery of everything which is likely to go wrong with every mower. Such a volume would be obsolete before its publication date. The hope was rather to formulate thought-patterns for consumers which will be useful in coping with most lawn mowers suffering a particular disorder.

Some important members of the lawn mower industry were extremely helpful in preparing this book. For the most part, the products they sell reflect their own personal integrity and a feeling that if the consumer buys a product, it ought to be safe first of all, and secondly it should be worth the price paid for it. Were I to mention their company names, you would know them immediately. These concerned individuals read and commented upon every word of the initial draft of this book. Their help was invaluable but some of them prefer that I do not now thank them by name. They are still a part of the lawn mower industry.

Legal commonsense makes it incumbent upon me also to state that, although every effort has been made to obtain the most accurate information and to present it in the most straightforward and precise manner, the publisher and the author can not assume responsibility for any kind of difficulties you may get into as a direct or indirect result of this book.

Brooklyn, New York
December 20, 1972

PREFACE TO REVISED EDITION

Authors seem to wax philosophical as their beloved books go back onto the presses. I'll keep my philosophies to myself, but I do feel angered enough about a couple of trends to share that with you.

In the 1973 edition's preface, I pointed out that there were 125,000 accidents involving lawn mowers every year. Now the figure is up to 160,000—that, after much hoopla about government-imposed safety standards and industry-volunteered safety standards! In fact, after reviewing what's been changed on lawn mowers over the past five years, I'm forced to conclude that *most* of the changes have been cosmetic. *Some* hardware has been tacked on to keep debris from being tossed at passersby, and *some* manufacturers have found ways to stop the whirling blades a bit sooner when the engine is turned off or the power cord disconnected.

While I'm happy that some changes for the better have taken place, isn't it sad that so little has been done in the area of lawn mower safety in five years? I'd keep that in mind, if I were you, when going out to invest in a new power mower. Instead of engaging in the national pastime of buying a new power mower every few years, spend a little more money for a well-built, safer mower. And then spend the few extra minutes a week required to keep it in good operating condition for five or ten years.

As for me, I'm keeping my hand-powered mower in good condition!

Madison, Wisconsin
November 15, 1977

HANDBOOK OF
LAWN MOWER
REPAIR

CHAPTER 1

Buying the Right Mower and Using It Properly

EVERY time you step into a store to buy a new lawn mower, you're in more quicksand than you may realize. Nobody effectively regulates the kind of safety equipment or durability that must be built into a lawn mower. Very few companies actually manufacture lawn mowers from the turf up, yet literally hundreds of companies sell products built in whole or in part by others.

Lawn mowers come in a cornucopia of sizes, shapes, types and colors. The proper one for you depends upon the variety of grass you have, the kind of use your lawn receives, the size yard you have to mow, and even whether or not children play on or near the turf.

THE ROTARY TYPE LAWN MOWER is the most popular today largely because of its price. There are few moving parts on the rotary mower beyond the engine itself. Most gasoline engines for small lawn mowers come from three major engine manufacturers. Consequently, a company needs almost no fabricating equipment to get into the business of assembling lawn mowers.

The engine on a rotary mower whirls a straight, sharpened blade at speeds in the vicinity of 3000 r.p.m., slicing off grass very much like a guillotine. Some machines employ more than one whirling blade.

Only the very tip of a rotary blade does the cutting, the outer two or three inches at most. The faster a blade spins, the better grass cutting job it can do. But beyond a certain speed, centripetal force would tear the blade apart. To balance safety with cutting efficiency, the industry standard is set at a compromise figure, a maximum of 19,000 feet per minute at the tip of the blade. A long blade spinning at 3000 r.p.m. will have a much higher speed (in feet per minute) at the tip than a short blade also rotating at 3000 r.p.m. Expressed in other units, 19,000 feet per minute is 216 miles per hour.

A 26 inch blade, for instance, will reach a tip speed of 19,000 feet per minute when the engine is turning at 2791 r.p.m. But an 18 inch blade will have to spin at 4032 r.p.m. to attain the same tip speed. You discovered this same principle on the merry-go-round in school; the farther from the center you stood, the faster you rode.

Very large power mowers generally use two or three blades of a manageable length rather than a single very long and unwieldy blade. Electric power mowers can make good use of blades under a foot long because electric motors easily spin at the 5000 r.p.m. needed to generate 19,000 feet per minute tip speed in a short blade. The typical inexpensive gasoline engine would literally fall apart at that speed. One 18 inch wide electric mower on the market uses two 9 inch blades. Thus the mower ends up roughly 18 inches wide but only 9 inches deep; a comparable machine with a single blade would be roughly 18 inches wide but would have to be 18 inches deep.

In order to attain speeds suitable for mowing grass, the engine on a rotary lawn mower must be operated close to full throttle almost continuously. If you drove the family car only at speeds around 100 m.p.h. you would expect to wear out most of the moving parts very quickly. That's what happens to the engine on many rotary mowers.

One other factor complicates the life expectancy picture for rotary mowers. They are very often sold with the hint that almost no maintenance is required. A shopper who *believes* such a sales pitch may soon find that a rotary mower going without good, regular maintenance is a short-lived piece of equipment. The average rotary mower going without good, regular maintenance seems to last *two years.*

An executive from a major engine supplier states off the record that his engines could barely be expected to last through 25 hours of the kind of beating they receive atop the usual rotary

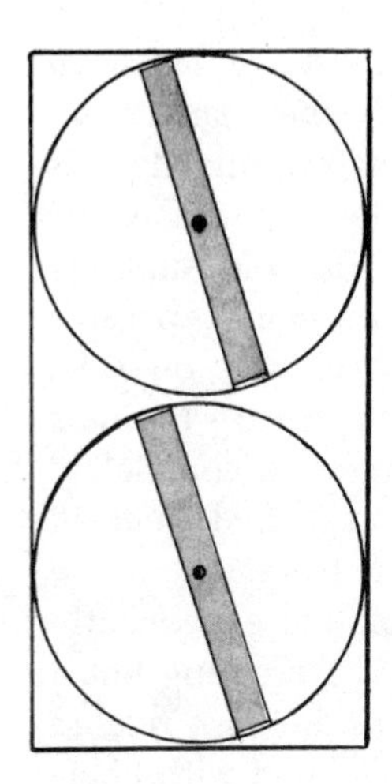

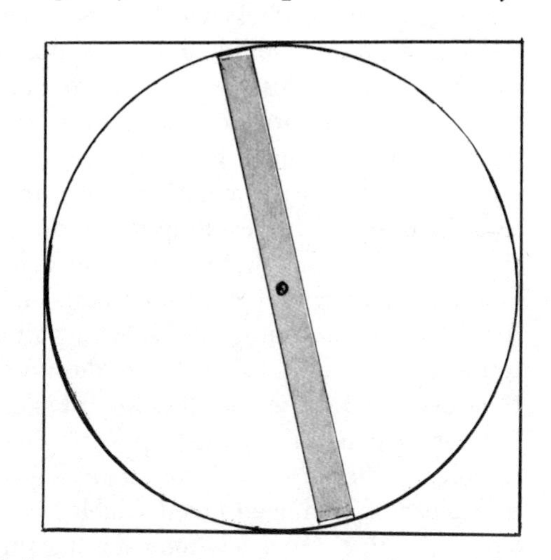

Illustration 1/1. Two Tandem Blades can cut a swath equal to a single longer blade, but require a drastically smaller mower housing. The right mower, for example, might be 18" x 18" whereas the one at left could be only 9" x 18".

lawn mower. There is no reason why even a rotary mower used frequently should not last *five years* if it is selected carefully, used carefully and maintained carefully.

A lawn mower body—or deck as it is called in the trade—should be made of iron. The most popular mowers sold today have decks made of aluminum or magnesium alloys, ostensibly because of *weight.* However, you'll find that a company's higher-priced mowers will more often than not come with iron decks. The total difference in lawn mower weight between light alloys and iron is approximately two pounds. The difference in life span for comparable mowers could easily be two years.

Surprisingly enough, the shape of a rotary mower deck is of

some real importance. If you keep your lawn trimmed fairly short and if you have a number of obstacles such as shrubs or trees or permanent outdoor furniture to mow around, you'll want to mow as close to them as possible. The alternative, of course, is going back around the edges with a hand trimmer.

The ideal shape for cutting close to obstacles would seem to be a perfectly round deck which hugs the outside tip of the blade. Such a covering would allow the operator to shove his lawn mower blade as close as possible to trees and other barriers on the right, left or front of the mower. A perfectly round deck is not especially strong, however, unless the deck manufacturer employs very sturdy metal and incorporates ribs and supports into the design. The resulting round deck is often more expensive than some of the more streamlined looking models on sale. Sensation mowers have become noted in the lawn care business for their rugged, round decks. As an alternative to a completely round deck, some manufacturers design a scoop-shaped front on their deck so it can come close to obstacles in the front.

Another structural feature appearing more and more frequently is called a *shroud.* On all too many machines it is nothing but a thin piece of enameled metal which covers the otherwise rugged looking engine. On a few mowers, the shroud is lined with fiberglas or other insulation and is shaped to absorb or deflect at least some of the rather hefty roar which a gasoline engine generates. You'll have to look underneath or unbolt the shroud to discover whether it is functional or decorative.

Since a steel sword whirling at 3000 r.p.m., or nearly 200 m.p.h., only inches away from your body is a potentially lethal weapon, safety must be of vital concern to you in considering any rotary lawn mower you might buy. One important safety feature is a trailing steel shield hinged onto the back of the rotary mower deck. Since the blade enclosure can be several inches off the ground, sizable objects are often tossed backwards at the lawn mower operator. This hinged shield guards against that likelihood.

There should also be a steel cover over the grass discharge

port in the lawn mower deck. It will lift out of the way when you attach a grass catching bag to the mower, but when the mower is used without the bag, the steel cover will help prevent glass, bolts, wire, stones, toy trucks and other potentially harmful objects from being hurled sideways through the deck's often very large opening.

As a rule of thumb, the best handle a lawn mower can have is the simplest. Frankly, it is hard for the average consumer to evaluate the strength of metal used in a lawn mower handle. Rust is a big villain responsible for shortening lawn mower life, however, so look for handles with as few rust-producing moisture traps as possible. Every extra bolt, attachment or decoration represents a potential location for rusting to begin.

The engine is the single most expensive part of a lawn mower. A worn out engine is also the most common reason for tossing out a machine. So give special attention to the engine's size, make and repairability.

Fully two-thirds of all small lawn mowers sold in North America include a Briggs & Stratton engine. The little Briggs engines have traditionally been the Rolls Royce of rural power systems. A farmer can't afford to buy a faulty engine if he expects it to pump water or drive a generator, and farmers in this country have traditionally bought Briggs & Stratton. However, Briggs & Stratton seems to be giving in to pressures for big quantities and low prices too. It would be very simple to say here, "Go out and buy a mower with a Briggs & Stratton engine of the proper size, take good care of it, and you'll get a maximum life-span." But you'll have to examine the model number on Briggs & Stratton engines because they also supply "special engines" for large lawn mower assemblers.

An executive at National Lawn Mowers remarked recently that "Briggs & Stratton has done a great service to the lawn mower industry with the engines and servicing they supply." But he went on to add, "I have my doubts about those special order engines though." Chapter four deals with repairing Briggs & Stratton engines and includes information about how you can identify a standard service engine from one of Briggs' lower priced, special order engines. If in doubt, ask the salesman if the engine crankshaft, which needs replacing all too frequently, is stocked by a local or regional supplier. On the special order engines, the crankshaft generally is also a special order item and can take weeks to obtain. After the salesman assures you that spare parts for this very mower are sitting on the shelf, ask him to put it in writing!

In the 19 to 22 inch rotary mower category, the engine should be at least 3.5 h.p. There are plenty of machines on the market with a 3 h.p. engine because it costs less. A mower with a 3 h.p. engine *might* be ok, but be especially skeptical of a mower featuring a 3 h.p. engine if the same assembler has a mower of the same size which offers the 3.5 h.p. engine as "an extra." It shouldn't be looked upon as an extra! Any engine has to work hard to maintain 3000 r.p.m. but if it has to work harder still because it is lacking in power, shortened life will result.

As part of a sales pitch, engine sizes are often quoted not in terms of horse power but in the cubic inches or cubic centimeters (c.c.) displacement of the cylinder. In truth, it actually is a more accurate way to represent engine size than in horsepower, but it seems doubtful that accuracy is the motive because displacement figures are quoted most often with very small engines. One well known brand, for instance, advertises that its self-propelled mower has a 101 c.c. engine but doesn't mention horsepower. For comparison, a Briggs & Stratton engine rated at 3.5 h.p. has almost 147 c.c. displacement. The smallest displacement which could be found for an engine listed by its manufacturer as carrying a 3.5 h.p. rating was 109 c.c. for a Lawn-Boy 2-cycle engine.

On rotary lawn mowers, it is the engine crankshaft which generally causes the most serious problems. The crankshaft takes power from the piston and carries it to the mower blade. If the blade is out of line so much as 1/16th of an inch or out of balance by a fraction of an ounce, the crankshaft is subjected to extreme vibration and bending forces. If the crankshaft bends, the lawn mower becomes virtually useless until a new one is installed. The replacement can cost you half the price of an entirely new engine or about one-fourth what you would pay for a new lawn mower.

Ideally a crankshaft should protrude from the engine only half an inch and be coupled to the blade with a soft key or shear pin. Unfortunately mower designers don't generally reach that goal. About the shortest exposed crankshaft length right now is 3/4 inch and some are as long as 2 inches. The more a crankshaft extends beyond the engine's bearings, the more likely it is to bend.

Unless you are persistent enough to examine the bottom side of a rotary mower to see for yourself how much of the crankshaft extends between the engine bearing and the blade, you'll have to rely upon the salesman to find out for you. You might make him put it in writing. Some manufacturers include a cone shaped crankshaft protector between the engine and the blade to act as an auxiliary, simple bearing. It does offer the shaft more support than no protection at all. One company advertises a five year guarantee on their mowers' crankshafts, but they don't qualify the statement in sales literature and the tiny user manual which accompanies their mowers makes no further mention of the generous offer.

One of the most interesting recent additions to the market place of lawn mowers has been the electric mower, rotary and reel. Noise is the major shortcoming of gasoline lawn mowers which electric ones have been able to overcome. Starting an electric mower isn't much of a problem either if your extension cords are strung together properly. Battery powered models tend to be too heavy and too expensive to be competitive for home use. That, too, may change.

The most common type electric motors found on rotary mowers are in the vicinity of 1-1/2 horsepower. In terms of power, the 1-1/2 h.p. electric motor does not compare unfavorably with a 3-1/2 h.p. gasoline engine as you might at first think. Both figures are *theoretical* ratings. The gasoline engine theoretical horsepower rating is about double its practical, useable power whereas the electric motor theoretical rating is much closer to its real and useable output.

Electric motors used on present lawn mowers not only have substantially less real power, they also create an entirely new problem. If a gasoline mower plows into grass which is too heavy for its power capacity, the engine stalls. Plow into the same over-load with an electric mower and the motor will try to keep turning, possibly burning itself out. For the present, therefore, electric lawn mowers are principally recommended for well established lawns. You can shove a gasoline mower through heavy weeds and long neglected grass, but you shouldn't attempt the same feat with an electric mower.

If you buy an electric lawn mower, you must also buy enough three-wire extension cords of the proper gauge to reach the farthest part of your yard. The cord should be bright, like red or orange, so you won't accidentally mow over it. Traditional yellow or grey cords will blend too well with the lawn at certain times of the year.

Some manufacturers may say that you can plug a three-wire cord into the ordinary two-wire outlet by using a cheap adapter. To do so, however, is inviting serious injury. The use of 3-wire-to-2-wire-adapters is tolerated in the United States but it has been outlawed in Canada because of its inherent danger. You are never entirely certain that the screw holding the outlet cover in place really is grounded as it should be. And that little pigtail can come unfastened from the grounding screw and go unnoticed until you suddenly receive an electric shock which should have passed through the grounding wire instead of your body. If your home doesn't have a suitable outlet for a three-pronged plug, install one or have it installed by an electrician or a handyman who knows what he's doing.

A lawn mower can weigh from 50 to 100 pounds. Consequently, many people are tempted to buy self-propelled rotary lawn mowers. In these models, however, the engine not only has to keep a blade whirling through grass at tremendous speed, it must also pull a relatively heavy load. There are also some serious questions about safety with self-propelled rotary mowers. Chapter 6 digs deeper into this problem.

REEL TYPE LAWN MOWERS have kept the grass clipped for generations. The hand lawn mower which lasted as long as the mortgage isn't an antique yet. It still is being manufactured but professional lawn care people who prefer a hand mower for trimming are the biggest customers.

The myth that rotary mowers can be run with practically no maintenance has been one major factor in the relatively short life-span of the rotaries. The notion that reel type mowers require extensive maintenance may have contributed substantially to their longer life. Apparently people who buy a reel mower expect to adjust the reels and perform routine maintenance on the engine. If the same kind of care were given rotaries, the rotary mowers which frequently have to be replaced after two years might well endure four or more years of use.

There are also important engineering reasons why reel type mowers outlast rotaries despite the fact that there are many times more moving parts which require some attention in a reel mower. One manufacturer of reel mowers, National Mower Company, says that it sets 10 years as the minimum working life that must be engineered into its product. Reel mowers use five, six, often seven or more revolving blades and one stationary blade to do their work. The rotating *reel blades* generate a shearing action against the stationary *cutter blade.*

Reel blades do not depend on brute force and speed for their cutting effect. A 2 h.p. engine going half-throttle can drive a self-propelled reel mower comparable in size to a rotary mower which requires a 3.5 h.p. engine racing close to full-throttle. Many reel mowers also use a 3.5 h.p. engine these days, and the resulting mower is much quieter than a rotary, creates less engine heat and destroys fewer piston rings, valves and crankshafts in the engine. Virtually every power reel mower made today is self-propelling.

A good reel mower generally will have six blades at a minimum. Some companies offer a five blade model apparently in an effort to keep down the cost, hoping to compete with rotaries on their own terms. Coarse grass under favorable conditions will cut well with a five-blade reel, but why limit yourself?

The actual number of blades is second in importance to the number of *cuts per foot* which the blades make. A quality reel mower should make no fewer than 10 cuts per foot. In other words, 10 blades should pass over the cutting bar for every one foot of forward travel the mower makes. Since cuts per foot is a figure that a salesman isn't likely to know unless the number is exceptionally high, don't hesitate to roll a showroom model across the floor, counting cuts as you push the mower for one carefully measured foot.

In theory, a mower with 5 blades which makes 11 cuts per foot should trim your lawn just as properly as a reel with 7 blades also making a total of 11 cuts per foot. The 5-blade reel would have to spin one-third faster and therefore could receive more wear.

If a reel is to maintain correct adjustment and if the blades are to stay sharp, the reel construction must be solid. A flimsy 7-blade reel would be no bargain compared to a rugged 6-blade model. One of the most rugged machines around advertises the size of its blades: 3/16″ thick by 1-1/4″ wide. Don't expect to find blades of that proportion on every mower in the supermarket but do keep the dimensions in mind. Well-built reels with blades 3/32″ thick and 3/4″ wide might be entirely adequate for careful home use.

At one time, the price of a rotary mower was as much as half that of a reel mower. Bad publicity caused by hundreds of thousands of bad accidents involving rotary mowers resulted in major and expensive alterations to the basic design of rotary mowers. Those changes, although still not entirely adequate, have made the price of rotaries and reels almost equal. And if safety standards get tougher still, the price of rotaries may equal or surpass the price of reel mowers. For your extra investment in a reel mower, you get quieter operation, longer engine life and an inherently safer piece of machinery.

The movement from hand powered mowers to gasoline powered mowers is extending now to mowers that not only power themselves but will also carry a grown man riding piggyback. Golf courses have for years used riding reel mowers for cutting their acres and acres of fairways. But greens to this day generally are mowed by hand.

Harry's Quarter Acre now needs a riding mower to stay in style it seems. Expert opinion on the trend toward miniature riding lawn mowers tends to be cynical, concluding that if walking behind a lawn mower for a short time once or twice a week is too strenuous, then your money would be better spent on Astro-Turf than on a small riding lawn mower.

Almost all small riding lawn mowers are rotary types. When you are sitting on one, you are only inches away from one or more sharp steel blades spinning nearly 200 m.p.h. If you must ride while you mow, shop carefully. Buy the riding mower which has the fattest tires because they'll be less likely to compact and furrow your soil. Buy the mower with the biggest engine available because it will be less likely to burn itself out or to chew up the transmission. Look for all the safety features you can find; in particular, there should be a switch on the seat which will stop the mower blades within seconds after you leap or fall off the mower.

Make sure that if your foot is removed from the clutch pedal, or is jerked off, the mower will stop—and not just the wheels. The blades must stop too.

If your lawn has some pretty hills and soft valleys, pay particular attention to the riding mower's center of gravity and traction. You don't want it to topple over or slide while you're mowing the hillside.

A short turning radius might be handy to have when maneuvering around trees, but if you turn too short on a hill which is too steep, the entire riding mower is apt to tip over. Unfortunately, matters such as center of gravity and turning radius are hard to assess in a showroom. It is important that you try out a demonstration model or visit a home owner who already bought a mower similar to the one you are considering.

While examining riding lawn mowers which do nothing but cut grass, you might also think about investing in a small tractor.

The garden sized tractors are several horsepower larger than riding mowers and cost significantly more, but they allow you to attach snow blowers, snow plows, garden plows, trash carts and other accessories. You can also couple tractors with the much safer reel type mowing equipment instead of the rotary blades built into most of the smaller riding mowers.

SHOPPING FOR A LAWN MOWER is not simple. There are something like 100 different lawn mower assemblers selling products under several hundred different brand names. *Consumer's Report* and *Consumer Bulletin* are of very limited help when it comes to lawn mowers. *Consumer's Report* goes for years with little more than a mention about lawn mowers. *Consumer Bulletin* may tackle two dozen different mowers by a dozen different companies, a fraction of the actual number of products you will encounter in shopping centers or super markets. *Consumer Bulletin's* listing specifies the engine manufacturer for only *some* mowers, a practice common throughout the lawn mower trade.

Some lawn mower assemblers specialize in putting together a varied line of mowers which chain stores offer under their own brand names. The manual which comes packed with such a mower doesn't mention any brand name, allowing the private label supplier to use the same manual with machines sold to several competing stores.

Because a product is a private label item does not mean it's bad. The G. W. Davis Corporation, for example, says that it supplies mowers for stores such as Kroger, Korvette and TopCo. Davis puts together some very fine mowers and some lower priced models as well. The trouble is, you have no way of knowing how much quality a store decided to offer in its lawn mower department short of carefully examining every major feature on each lawn mower displayed. You may also have trouble obtaining replacement parts unless the store and the private label manufacturer are both unusually conscientious.

Generally it is worthwhile to spend a few dollars extra to buy a company's top-line mower instead of the economy model.

The economy in the economy models often ends at the sales room. Lower-priced mowers can mean a compromise on safety features and prematurely worn parts. Moving from the economy model to the top-line model can typically mean going from a 3 h.p. engine to 3.5 or 4 h.p., from a thin aluminum deck to a hefty iron one, from five blade reels to six, from a machine with minimum safety features to one with the "optionals" which ought to be standard.

Moving upwards in price for features which bring safety and durability is a bargain. Higher prices because of chrome, worthless shrouds and racing stripes is a matter of taste, not economy.

Some optional extras might be quite worthwhile, however. If you experience difficulty pulling a starter rope, it is possible to find an engine with a simple wind-up starter or electric one. If a shroud is designed to reduce engine noise, it ceases to be just decoration and could represent a worthwhile extra. In some mowers, the exhaust system is twisted so the muffler blows exhaust and sound beneath the rotary mower deck, reducing noise a bit further.

If your lawn mower breaks down on July 4th, will you be able to find replacement parts or a repair man on July 5th? And if you keep your mower running for 10 years, will spare parts still be available? Of all the sales and user literature examined for this book, only Yard-Man said in writing in their user manual that spare parts would be available for 10 years.

Before finally deciding on a particular lawn mower, make sure that replacement parts and service are available locally both for the mower itself and for the engine. Does the store which sells you the lawn mower also sell replacement blades, bolts, belts, chains, spark plugs, points? If they don't, who does? Do you get a written guarantee that you'll be able to buy spare parts—any spare parts—for 10 years?

SAFETY IS SOMETHING WHICH EVERY LAWN MOWER USER SHOULD BE VITALLY CONCERNED WITH. Very few lawn mower assemblers seem to be. Over 100,000 people are injured annually in power lawn mower accidents according to the National Commission on Product Safety. Three-quarters of those accidents are fractures, amputations and lacerations caused by whirling blades on a rotary mower.

Any family with small children, or anyone living in a neighborhood with small children, should consider very carefully what kind of mower it ought to own. If all of the factors about safety are weighed, if all of the factors about noise and durability are weighed, it is difficult to understand why anybody would tolerate a rotary lawn mower in the vicinity of children. It isn't always the man or woman behind the mower who gets injured. A piece of wire or a nail or a bolt or a piece of broken glass thrown from a rotary mower can have the striking power of a .22 caliber bullet!

Why doesn't Congress do something about lawn mower safety standards as it finally began doing with auto safety? That's a question you could very well ask your local congressman! The House Commerce Committee went 10 years without considering lawn mower safety.

The lawn mower industry, through its Outdoor Power Equipment Institute, Inc. (OPEI), has made some effort at policing the quality of lawn mowers. However, its members come almost exclusively from the ranks of the lawn mower industry. Some manufacturers of better lawn mowing equipment will not join the OPEI. The organization itself feels that, since its membership includes companies in virtually all phases of lawn mower assembling, it should not comment on the relative merits of various types of products.

The OPEI sponsored a project at the American National Standards Institute, Inc. (ANSI) which developed the current standards for minimal safety throughout much of the lawn mower industry. Despite their similar name, there is no connection between this private organization and the government's National Bureau of Standards.

The ASNI says, in introducing its publication on lawn mower safety standards, "An American National Standard implies a concensus of those substantially concerned with its scope and provisions. An American National Standard is intended as a guide to aid the manufacturer, the consumer, and the general public. The existence of an American National Standard does not in any respect preclude anyone, whether he has approved the standard or not, from manufacturing, marketing, purchasing, or using products, processes, or procedures not conforming to the standard. . . ." The 19 page publication costs as much as many full length hard cover books.

After the ANSI standard on lawn mower safety was issued, lawn mower assemblers began sticking decals on practically every one of their mowers, proclaiming that it complied with the ANSI standards. However, the National Commission on Product Safety reported that when it tested 200 mowers bearing the safety decal, 55 did not meet the ANSI standard at all.

Current practice in the lawn mower industry is to have an outside laboratory—unnamed—test lawn mowers to determine whether new models do conform to ANSI safety specifications. That is not to say every individual lawn mower sold by the company will meet the standard, only that a sample of a new model met test standards.

The committee which decided on the consensus which became the ANSI standards included *two* representatives of safety related organizations, *two* insurance associations, *one* lawn mower user organization (the American Cemetery Association), and *four* representatives from industry trade associations.

Lawn mowers bearing an ANSI safety decal probably are safer than similar mowers sold before the standards were published. That is not to say or even to imply that mowers today are as safe as they could be and should be. There could be many safe lawn mowers sold without the ANSI decal.

Specification 3.3 in the ANSI standards say that a mower must have an audible or visible signal if the power source is so quiet that the user wouldn't otherwise know the blades were rotating. When the power is turned off, however, the standard does not insist that the signal must continue until the blades actually come to rest. Electric motors make very little noise and if the signal quit when somebody kicked out the cord, a user might not know if the blades had coasted to a stop. On gasoline powered rotary mowers with belt driven blades, it is entirely possible for the engine to stop chugging 10 or 15 seconds before the blades silently coast to a stop.

The next specification, 3.4, says that blades should coast for no longer than 15 seconds after power has been disconnected. Look at your watch right now. Fifteen seconds is a long, long time!

Printed lists of safety rules are frankly of questionable value. If the list is long enough to be comprehensive, it's too long for most people to remember. And if it's very short and readable, something has been left out. The ANSI, via its concensus method, came up with a condensed list of rules for lawn mower users which is widely reprinted in manuals which accompany new lawn mowers. The list is reprinted in this book's appendix. It wouldn't be a bad idea at all if you hang up the safety warning wherever the lawn mower gets stored. You might pause for a minute and reread the list occasionally that way.

THE SIZE OF YOUR NEW LAWN MOWER has been the subject of much promotional literature. You can buy 18-inch mowers, 20-inch mowers, 21-inch mowers, 22-inch mowers and on up the line. The same catalog from the same lawn mower assembler can list 20, 21 and 22 inch mowers as if the difference of a few inches in blade length was highly significant.

A few minutes with a pencil and paper can demonstrate just how unimportant a few inches of blade length can be. Let's assume it takes the same length of time to dodge trees and shrubs with an 18-inch mower as it does with a 24-inch mower. Let's also assume that the two mowers both move along at a leisurely 2.5 m.p.h. Allow a casual 3 seconds for turning the mower at the end of every strip of cutting. Consider that the hypothetical lawn is a solid 100 feet long and 50 feet wide, without shrubs, trees or a house in the way.

With an 18-inch mower you would have to make 34 trips up or back the 100 foot length. Behind a 24-inch mower you would make 25 trips. You would have to turn the 18-inch mower 33 times and the larger model 24 times. For the 18-inch mower, the theoretical elapsed time is 17.1 minutes and 12.6 minutes for the

24-inch model. About 4 minutes and 30 seconds is all the time you might save by using an unusually large mower instead of an unusually small one on an average sized backyard.

In this same theoretical set-up, a 20-inch mower takes 15.1 minutes and a 22-inch model 14.0 minutes. If you can spare 1.1 minutes a week, you'll probably save a few dollars by buying the

Illustration 1/2A and 1/2B. Lawn Mowing "Road Maps" which keep the discharge port of a rotary lawn mower pointed away from the house (or play area) as much as possible.

smaller of these two popular sized mowers. Often there is also a difference of several pounds to push around.

THE PATTERN WHICH YOU CONSCIOUSLY OR UNCONSCIOUSLY ADOPT WHEN MOWING THE FAMILY LAWN is more important than you might think. Sticking to any one pattern is undesirable. If you walk only back and forth every time you mow the lawn, many strains of turf will begin forming bumps and ridges *perpendicular* to your usual direction of cut. And if you have a riding mower with wheels that are not especially broad, the weight on those tires can compact the soil into ruts *parallel* to your normal direction of mowing.

Every mower company which suggests a mowing pattern seems

to have a different approach. Some prefer diagonal swathes, some a straight up and back, others a circular system. Your best approach is to alternate patterns as much as your own particular lawn allows.

Safety also should be kept in mind when selecting mowing patterns. With a rotary mower, you should work out a plan which will keep the discharge side of the mower from being aimed at close range toward the house or play areas. Drawings number 1/2a and 1/2b show possible plans.

THERE ARE THREE OR FOUR PRINCIPAL VARIETIES OF GRASS SEEDS, and within each variety what seems like dozens of brand names. Certain types of grass fare better with reel mowers, others with rotary type mowers.

The lawn industry is overgrown with institutes and foundations and commissions, few of which have any real objective other than promoting a particular product for profit. If you haven't yet selected the lawn and lawn mower combination which is going to grace your house, estate, golf course or park, tred lightly on the claims of experts bearing credentials from an institute or foundation. Even academic credentials should be viewed skeptically since at least two Ph.D.'s, who are among the more widely quoted agronomists in the turf field, are on the staff of two major lawn mower companies.

Judging by the varied opinions in print, there is no ideal height for mowing grass. Factors which should influence the height you select for cutting your lawn include the time of year, the type of grass seed, if the lawn receives hard or casual use, and whether or not weeds are an immediate problem.

Bluegrass, for example, is one major and popular specie of grass seed. Merion bluegrass, however, is an industry brand name for a sub-specie of bluegrass discovered, so the story goes, at the Merion Golf Club near Ardmore, Pennsylvania. There is even a public relations office on Park Avenue in New York, The Merion Bluegrass Association, which promotes the use of Merion Kentucky Bluegrass seed. Gracing some of the Merion literature is a seal, "Approved by Turf Research Foundation" which incidentally shares the same address and telephone number as the Merion Bluegrass Association.

Professor of Horticulture at Kansas State University, Ray A. Keen, writes that only the reel type of lawn mower can produce the very finest of turf because it is able to shear off the grass more uniformly. He points out that the reel type mower is used on golf greens, bowling greens and similar fine turf areas. Professor Keen finds the rotary mower excellent for what he calls less perfect lawns and in taller grasses.

At Purdue University, William H. Daniel from the Department of Agronomy writes that bluegrasses are well suited to rotary mowers. Bentgrass, Zoysia and Bermuda grass varieties sometimes become stringy and then might be pulled up to create patches of brown when used with rotary mowers. Balancing his comments, however, Mr. Daniel feels that since rotary lawn mowers cannot be set to mow as short as the reel types, the introduction of rotary mowers on a large scale may actually have resulted in overall superior home lawns. He feels that the average home owner may try to mow his grass too short.

To meet the varying demands of lawn height, the Kansas State University of Agriculture and Applied Science recommends a lawn mower which can be adjusted from 1/2 inch to 3 inch cutting levels. Few mowers on the market today can be adjusted for cutting both that high and that low. It is virtually impossible to find a rotary mower able to cut grass to a height of 1/2 inch.

Close cutting in the spring can encourage grass to spread out, to create a denser turf and give less room for weeds to take root. Later in the year, when weeds may be a problem, let the grass grow longer to shade the young weeds and keep them from growing as vigorously as they would if given a chance to share the sun with your grass.

If tall grass is suddenly cut short, the green leaf area is reduced so much that grass can go into shock or at the very least grow with considerably less vigor. A good rule of thumb ascribed to by most agronomists and lawn experts is not to cut off more than one-third the height of your grass at one cutting. A lawn which grew to three inches since the last cutting should therefore not be trimmed closer than two inches. After a few days growing time, you can trim the same lawn closer than two inches if you want. Closely trimmed lawns have to be mowed more frequently during heavy growing periods. If you want to maintain a lawn at a cut of 1/2 inch, it should be mowed every time the grass has reached a height of 3/4 inch to avoid cutting off more than one-third during a single mowing. A lawn kept normally at 1-1/2 inches can safely grow 1/2 inch between cuttings.

The warm-season or creeping grasses generally do better with close cutting. Bermudagrass, centipedegrass, zoysia, carpetgrass, St. Augustinegrass and bentgrass come under that category. They generally look best and grow best when kept at a height of one inch or less.

Bluegrass, rye and fescue receive a recommended cutting height from 1 inch up to 2 inches, but a general consensus of lawn care experts seems to favor 1-1/2 to 2 inches. The newer strains of bluegrass, Merion, Prato, Windsor and Fylking, can do well at a 1 inch cut.

With so many industry experts and academic experts and self-proclaimed experts offering diverse advice, pamphlets and articles on the merits and demerits of grass cutters, this book hopefully can help the consumer pick his way through the tangle.

CHAPTER 2

Routine Maintenance: Pill for Long Mower Life

YOUR lawn mower should last at least five years. Ten years is not too much to ask. This is true no matter how the mower's engine, deck, blade, handle, pulleys or transmission are designed, and no matter how little information may be included in the user manual.

The secret to long life for your lawn mower is based almost entirely upon regular but simple maintenance. In order to hit the 10 year life span, a figure which will make most lawn mower salesmen shudder, you must not only properly oil, grease, adjust and clean the mower, you will have to replace some of the moving parts. Consequently, spare parts must be available from the assembler of the particular lawn mower you choose to gamble on. In the case of the engine, parts must be available from the engine manufacturer. Let this be a warning to you when selecting a lawn mower.

THE PROPER KIND OF GASOLINE, OIL, GREASE AND EVEN STORAGE AREA is vital for long engine and lawn mower life. These topics receive precious little discussion in user manuals written to accompany new lawn mowers.

There *is* a significant difference in gasolines, and not just the well known distinction between "regular" and "high-test." The engines built for lawn mowers are designed to run on "regular" gasoline. Briggs & Stratton, for example, designs its engines to do well on fuel with an octane rating of 85 or higher. All "regular" grades of gasoline marketed now in the United States meet that standard.

Lead, which has been a traditional part of almost all gasoline sold for use in cars, should be regarded by lawn mower users as a bad impurity. Aside from polluting the air you breathe, lead in gasoline also impairs the performance of spark plugs, valves and piston rings. It contributes to larger, tougher combustion deposits inside your engine's cylinder than does no-lead or low-lead "regular" gasoline. Do not, as was suggested in literature from one lawn mower assembler, use "white gasoline."

Gasoline companies have for years been adjusting the content of their gasoline for the season and for the climate in your particular part of the country. For winter sales, the fuel is blended with light, highly volatile chemicals to aid in cold-weather starting. During the summer, the highly volatile additives are cut back and chemicals which prevent boiling and vapor locks are added instead.

The changing gasoline blend, so useful for your car, can also help your lawn mower engine which faces some of the same seasonal difficulties as your car engine. To get the most benefit possible from the changing seasonal blends of fuel, store as little gasoline as you find practical. Such a practice also makes for a safer garage.

Under no circumstances should you use gasoline which has been stored from one lawn mowing season to the next. Gasoline gets stale too. Oxidation in gasoline results in sticky deposits of gum or varnish which settle particularly on intake valves. Keep gasoline in a cool place to retard oxidation. Keep the fuel tank either completely full or completely empty whenever the mower is not in use since either condition will minimize the amount of oxidizing that can occur.

OIL COMES IN SEVERAL GRADES. Each grade represents a particular quality. To understand *why* your engine deserves the best quality of oil around, you must appreciate what oil does in your air-cooled lawn mower engine.

Lubrication is the most obvious use for oil. The pistons, crankshaft, valve stems, cams, gears and other moving parts move at a very high speed and with very small clearances between them. Without a thin film of good lubrication, friction would very quickly wear away too much of the metal. Since there is no water inside your air cooled engine, the oil also helps to cool the heated metal parts. Oil also acts as a seal between the cylinder wall and the piston rings. That oil seal helps hold the air-gasoline mixture above the piston during the compression stroke and prevents the burning gasses from escaping past the piston during the power stroke.

Oil for your air cooled engine must have good lubrication ability in addition to containing detergents and oxidation inhibitors. Detergents help scrub away combustion deposits which otherwise would accumulate and interfere with the piston rings and valves. For every gallon of gasoline burned in your engine, one gallon of water is produced by the combustion. Under normal operating temperatures, the water escapes as steam in the exhaust. However when the engine is cold, some of the water condenses on the cold cylinder walls and drops into the oil pool in the crankcase. Water droplets combine with other impurities to form acids or sludge or other damaging by-products. The oil in your engine has to be able to function without breaking down in the presence of these acids.

At the high temperatures of a gasoline engine, oil can char or oxidize in much the same way that cooking oil does on a hot frying pan if you forget to toss in the pancake batter in time. An oil which is unable to withstand 5000°F. or more of heat, can form varnish and charry deposits.

The American Petroleum Institute rates various lubricants according to the kind of service they can perform. You should select a detergent oil having a classification of "SC" or "SD," which means it will perform well under severe operating conditions. The older classification by the A.P.I. for this same oil was "MS". Although "MS" is practically identical to "SC" or "SD" in its qualities, a dealer still stocking oil with the old "MS" markings may not have a big turnover in his stock. If you are going to pay good money for a can of oil, you may as well get a good can of oil.

Aside from a quality rating, oils are also available in various viscosities. The thicker oils hold their lubricating powers better during high temperature operation and thinner oils allow for easier starting and faster lubrication during cold weather. Oil is commonly available with ratings from SAE 5, a very thin oil, to SAE 40, a very thick oil.

For summer operation, most engine manufacturers recommend SAE 30. If you have trouble finding SAE 30 oil, an oil classified as SAE 10W/30 or SAE 10W/40 is your next best choice. When the temperature drops below 40°F., if your mower engine does winter time duty on a snow blower for instance, SAE 5W/20 or SAE 5W/30 are the first and second most preferred choices, with SAE 10W/30 being next best. When the thermometer drops below 0°F., Briggs & Stratton recommends that you use SAE 10W or SAE 10W/30 oil diluted 10% with kerosene. This is 3-1/2 ounces of kerosene for one quart of motor oil.

Routine maintenance logically falls into three fairly distinct categories: (1) before or after every use; (2) periodic adjustment and lubrication; and (3) season-to-season care. In the Appendix, each of these three maintenance schedules has been reduced to a simple check list which you can cut out and mount under plastic in your basement, garage or storage shed.

EVERY USE: BEFORE MOWING

1/ *Check the level of both gasoline and oil.* You should have filled the gas tank at the end of your last mowing. If the level has dropped appreciably (assuming the mower isn't exposed to so much heat the gasoline evaporates), there may be leaks somewhere along the line. The oil level should be at "Full" on the dipstick. Never fill the crankcase with oil above the "Full" mark or serious consequences can befall many moving parts. If oil must be added frequently, begin to suspect some mechanical problem or some quirk in the way you use the mower which may be overloading or overheating the engine.

2/ *Look for loose parts on the engine.*

a. The air cleaner.

b. Oil drain plug which is located underneath a rotary mower. (Disconnect the spark plug wire in that case.)

c. Mounting bolts which hold the engine onto the lawn mower.

d. Spark plug wire.

e. Fuel line connections.

f. Gas tank mounting brackets.

g. Throttle, choke springs and levers on carburetor.

3/ *Look for loose parts on the mower itself.* This is especially important on rotary mowers.

a. Controls, particularly ones linking the mower handle and the engine. Pay careful attention to items like the clutch or brake or "STOP" control. You may want to use them in a hurry should an emergency arise.

b. Drive mechanism. On some smaller machines this is simply a friction or sprocketed drive against a wheel. In others, it can be a pulley or chain drive.

c. On riding mowers, check the seat support bolts. This is an essential item to inspect for safety. They are subjected to a great deal of vibration and abuse. Should one or more of them work loose while you are chugging across the lawn, you could topple into the moving mechanism.

4/ *Look for leaks.* Just before pulling your mower out of storage, shove it forward or backward far enough that you can examine the floor or ground where it had been laying in wait all week. Carefully analyze any new wet spots to discover if they are caused by oil or gasoline. Rub your finger into the wet spot; if it smells like gasoline, that's your answer. If not, it might be oil. Judging by where the mower has been sitting, try to determine from what part of the mower the oil leaked. Likely sources include the oil drain plug, the oil filler plug, the oil pan or sump, bearings around either end of the crankshaft, power take-off bearing or air cleaner on older models which contain a pool of oil.

5/ *Oil all of the principal moving parts.* This is done *after* you have checked for leaks. Otherwise drops of excess oil from this part of your maintenance could confuse you when looking for leaks.

A comprehensive list of parts needing oil before every use ought to accompany every lawn mower. Unfortunately this is

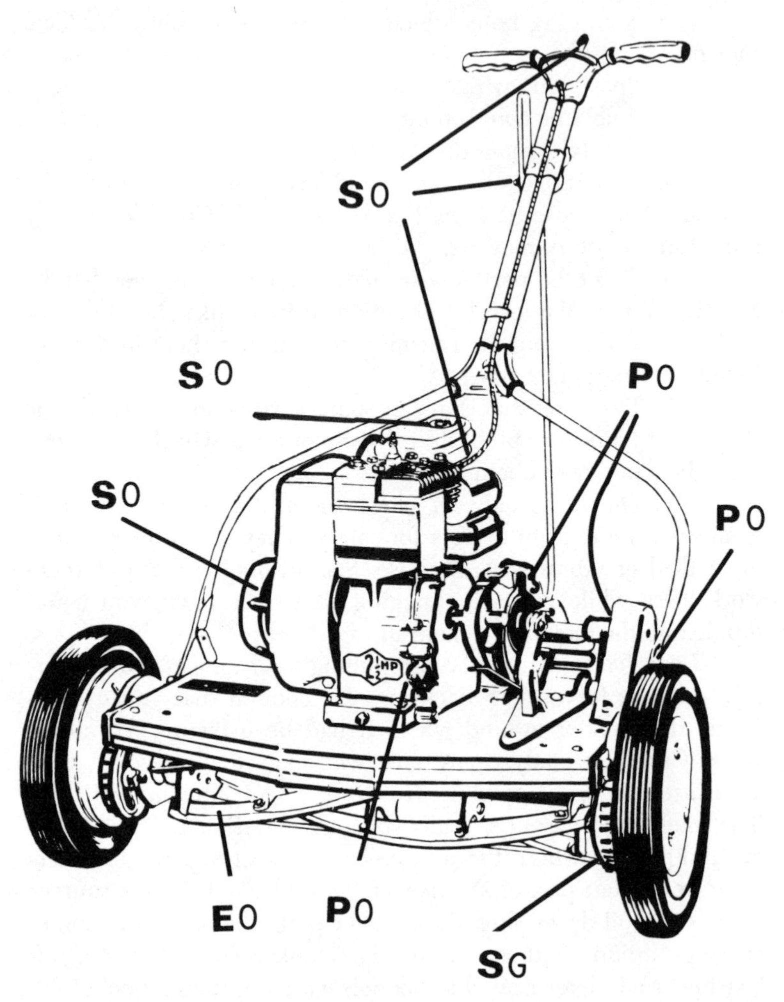

Illustration 2/1A and 2/1B. General location of parts requiring regular lubrication. G = Grease, O = Oil, S = Seasonal, P = Periodic, E = Every Use.

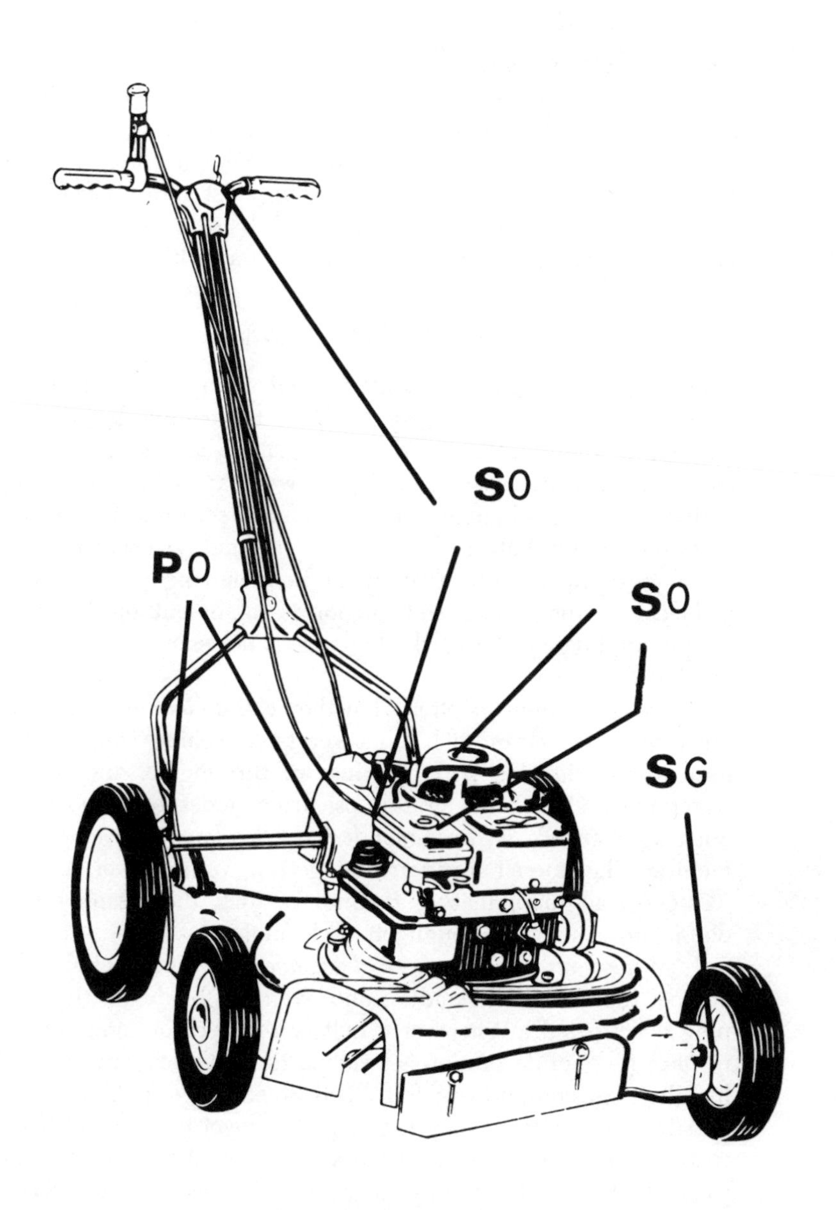
SO
PO
SO
SG

seldom provided. Drawings number 2/1a and 2/1b show some typical moving parts which benefit from regular lubrication. This illustration is only a guide, however, because mowers, of course, come in a rainbow of sizes, shapes and complexities. On reel mowers, a thin film of oil or silicone lubricant sprayed from an aerosol can onto the cutting bar and each blade on the reel both improves cutting performance and insures longer life.

EVERY USE: AFTER MOWING

1/ *Remove all grass, leaves, dirt and other debris from the mower, engine and moving parts.* Since the engine depends on air to cool the cylinder head, accumulations of dried grass in cooling areas can cause the engine to overheat. A further danger is that heat from an engine can occasionally set dried leaves and grass on fire. On both reel and rotary mowers, you will have to clean grass away from areas on or near moving parts, so disconnect the spark plug and anchor it safely out of the way. Simply unplugging the spark plug wire is not enough; it tends to flop back in place.

A few rotary mowers offer what they call *a cleaning port.* You are directed to clean such a mower by attaching your garden hose to the cleaning port, starting up the engine, and letting water power knock the grass loose from underneath the deck. Such a procedure is *recommended by the lawn mower* manufacturer who doesn't mention that getting your mower soaking wet can lead to premature rusting. Besides, the cleaning port doesn't always do a thorough job of cleaning.

2/ *Wipe your entire mower clean and dry.* This includes the engine, the deck, the handles, and even the blade (with spark plug deactivated.) Wherever a bolt or the deck or some attachment or piece of decoration is fastened to the typical lawn mower handle, corrosion and rust find ideal breeding grounds. It is impossible, of course, for any rag or paper towel to wipe all of the rust-causing moisture away from such traps. If you really want to get 10 years of life from your mower, the ideal way to prevent rust from eating through all the knick-knacks on your handle is to spray it gently with oil after every use. The same treatment is recommended for any other water trap on the deck or superstructure of your mower.

3/ *Fill the gasoline tank.* Save this for last, so the engine has a chance to cool down to a safe temperature.

PERIODIC MAINTENANCE

Home owners who use their mower at least 10 hours in a summer month ought to run through the periodic maintenance check list at least once or twice during every summer. Professional users, who can top 30 hours of wear every week, ought to consider periodic maintenance part of a weekly routine. Times given in (parentheses) are for the benefit of professionals or other heavy users.

1/ *Charge the battery.* (Every day or two for professional users.) Most rechargeable batteries on smaller lawn mowers can be brought back up to full charge by a full day on the small chargers which come with the mower. No harm generally will be done—unless instructions specifically advise otherwise—if nickel-cadmium batteries are left permanently on the charger. Why not just pull the battery out of its clip after every use and hook it up to the charger until the next grass cutting siege? Pros who own a mower which does not recharge the battery during use, will want to own at least two such batteries and switch them around once or twice during the week.

2/ *Remove grass and other debris from under the engine blower housing* or any other kind of decoration the lawn mower assembler built onto the top of your model. (Weekly for professional users.) Illustration number 2/2 shows the critical areas to keep clean, the cooling fins built into the block and head and the flywheel blower blades. The shrowd itself, if one exists on your mower, should be removed during this thorough cleaning.

3/ *Change the oil.* (Weekly for professional users.) The engine must be warm before all of the old oil will drain out. If you do this after cutting the grass, the matter is simplified. If not, let the engine run for 5 or 10 minutes before opening the oil drain plug. Clean away dirt and grass from the area near the drain plug before twisting it open. The used oil is likely to be pretty messy. Have a flat can on hand to catch the old oil.

On rotary mowers, you'll want to sharpen the blade at this time anyway, so why not remove it? Since the oil plug often is under the deck, having the blade out of your way will make draining the oil a lot simpler. (Deactivate the spark plug first.)

Illustration 2/2.

Before twisting off the oil filler plug or cap, clean the area well. Then add the best quality of detergent oil you can find, making sure it is of the proper classification and type.

4/ *Clean the air filter.* (Weekly for pros.) If your mower is used to dispose of dried leaves or if it's been a dusty summer, non-professionals ought to clean the air filter at least once a month. There are several different types of air cleaners and each has its own peculiar preference when it comes to cleaning.

The *oil foam air cleaner* is found on most smaller Briggs & Stratton engines today and is typically located right above the gasoline tank. Unscrew the wing nut and lift off the cover. Lift out the foam cleaning element inside and pry it *downwards* off the cup inside. Drawing number 2/3 shows the cleaning procedure. To reinstall, reverse the above instructions for removing. Make sure that the wing nut is tightened very firmly because vibration, otherwise, could loosen it with relative ease.

A *dry element air cleaner* is common on larger Briggs & Stratton engines and is standard for the engines of some other manufacturers. After removing the element from its housing, knock

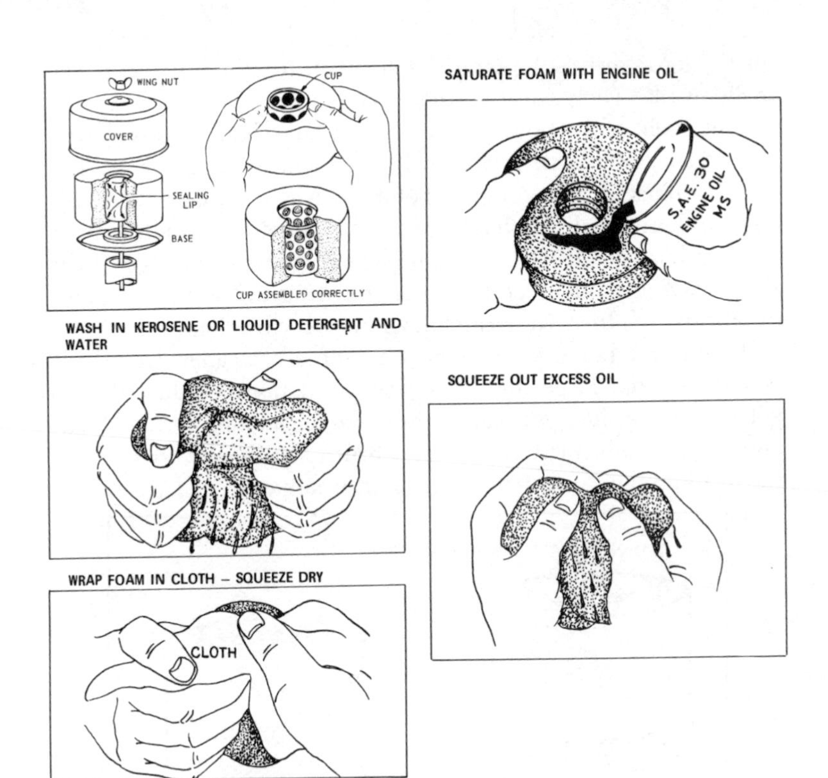

Illustration 2/3. Step-by-Step Procedure for cleaning an *Oil Foam Air Cleaner.*

out the loose dirt by tapping the element firmly on a workbench or sidewalk. Then wash it thoroughly with a non-sudsing detergent such as the powdered or liquid household cleaning aids. Most dish washing detergents will give you too many suds. The element must dry thoroughly before you can replace it. A half hour in direct sunlight should dry the element. As an alternative, you can blow air from a vacuum cleaner through the side of the element. Again, make certain that the wing nut is turned down tightly enough to resist normal lawn mower vibration.

If you clean a dry element type of air cleaner frequently enough so that large and sticky accumulations of dust do not get trapped inside, you can simply use a vacuum cleaner to do an adequate job. First use a small brush attachment to suck out the dirt and then switch to a forced air treatment.

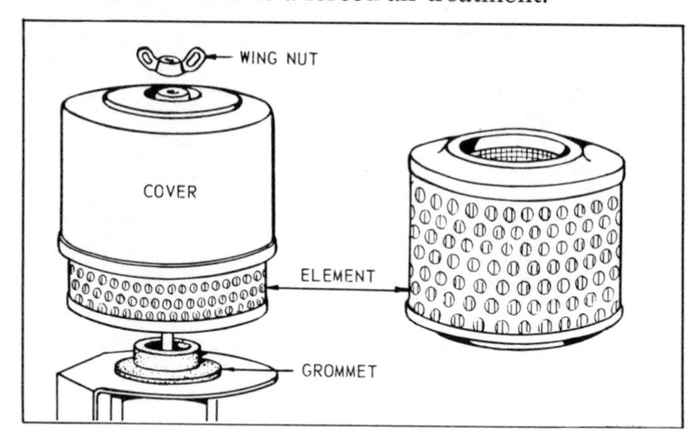

Illustration 2/4. A Dry Element Air Cleaner (metal, not paper element) is washed in non-sudsing detergents.

You may still encounter some *oil bath air cleaners.* Screw off the top cover cautiously because there is a puddle of oil inside the cleaner. Pull out the element and clean it thoroughly with kerosene or solvent. Let the element drip dry. Pour out the old oil from the bowl and replace to the indicated level with your normal grade of clean engine oil. Reassemble carefully so the new oil bath does not spill out.

5/ *Check and clean ignition points.* (Monthly for pros.) Larger

Briggs & Stratton and many other makes of engines have the ignition points located so they can be checked by simply removing a protective cover. Smaller engines, however, have the points behind the flywheel; in that case, you may decide not to pull the flywheel more than once a year. Unless your mower receives heavy use, it seems unnecessary to remove a flywheel frequently. If the engine suffers from ignition trouble, however, Chapter 3 will tell you how to locate the disorder. Chapters 4 and 5 will lead you through procedures for remedying it. Exploded drawings at the start of Chapter 3 will help you identify your engine and the location of ignition breaker points.

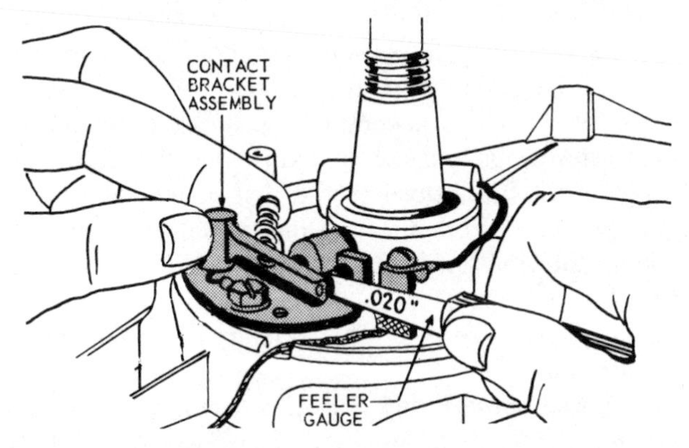

Illustration 2/5. Correctly setting the ignition gap on a Briggs and Stratton engine. The .020" gap varies in different machines.

Once your ignition system has been uncovered, the two contact elements on the points should look clean and smooth. Oil can be carefully wiped away with a clean rag dipped in solvent or kerosene. A feelers gauge is used to check that the points are set the proper distance apart, .020" for most lawn mower engines. If some of the preceding terms are unfamiliar, don't despair. They will be described thoroughly in subsequent chapters. The terminology and procedures are quite easy to pick up!

6/ *Clean the spark plug and set it to the proper gap.* (Monthly for pros.) Clean the area around the spark plug hole before removing the plug. The business end of it should be dull but not overly dirty. If the two elements have a crumbly appearance or are choked by deposits of carbon, first wipe them clean and then, if necessary, gently use a wire brush. If they clean up easily, check the gap and reinstall the plug. If they do not, buy a new spark plug and check that its gap is set properly before installing it.

Briggs & Stratton lists .030″ as the proper gap for its spark plug. The recommended plug gap for other engines varies all the way from .020″ to .035″.

7/ *Clean away carbon from inside the cylinder.* This ought to be done monthly for professional users and at the start or end of every season for others. It is necessary to remove the cylinder head for this relatively important operation. That procedure is dealt with at length in Chapter 4. The carbon, lead, varnish and gum deposits which accumulate in and around the pistons, cylinders, piston rings and valves if this procedure is overlooked, will probably not be noticed in the first couple of years of engine operation. But from then on out, you'll no longer be able to overlook the foreign matter trapped inside your cylinder head. Overlooking this bit of maintenance could very well be one of the major physical reasons why so many lawn mowers have so limited a life span.

8/ *Check the pulleys, belts, chains, sprockets and clutches.* (Weekly for professional use.) Tighten up all the belts and chains to their proper tension. If a belt is starting to show signs of wear, order a new one which you then can replace at a time of relative leisure. The same warning applies to any worn or damaged links in chain drives. After a sprocket is cleaned, check it for cracks or badly worn teeth. If it is showing signs of wear, order a new one. At this same time, you'll be able to take care of all necessary lubrication in these areas with oil or grease as recommended by the guide lines in drawing number 2/1.

9/ *Oil and grease all moving parts.* (Weekly for pros.) Consult your own mower's user manual and drawing number 2/1 for guidance. All of the points marked with E (every use) and P (periodic maintenance) in drawing 2/1 should be treated to oil or grease at this time.

10/ *Sharpen and balance the rotary blade, or lap the reel if necessary.* (Weekly for professional use.) Chapter 6 goes into blade care on rotary mowers. Chapter 7 covers reel and cutter bar care for reel type mowers.

SEASONAL
AT THE END OF EVERY SEASON

1/ *Drain the gas tank.* If you plan ahead carefully, you'll end up the last day of the grass cutting season with no more than a splash or two of gasoline left in the tank. If your engine has a separate fuel line (if the carburetor *is not mounted* directly atop the tank), disconnect the line as close to the tank as possible. Be sure that the engine has cooled down first. Carefully let the gasoline drain into some vessel. Add fresh, clean gasoline to the tank and slosh it about vigorously by shaking the engine, tank or mower while allowing the gasoline to drain out.

Blow air from a vacuum cleaner into the tank and through the fuel line. (Never suck gasoline vapors *into* a vacuum cleaner. A spark in the electric motor could cause an explosion.) The purpose of this procedure is to remove all gasoline from the tank and line so oxidation over the winter will not deposit gum inside the carburetor openings and diaphragms. Engines featuring carburetor and tank as one integral package must be emptied manually. As a finale to getting fuel systems ready for the off-season, the engine is started and allowed to run until all of the gasoline in the carburetor is consumed.

2/ *Change the oil.* You don't want to leave the moisture and acids caught by the old oil in the engine through the winter.

3/ *Remove the spark plug and pour a tablespoon of clean lawn mower engine oil through the spark plug hole.* Force the piston up and down several times by rotating the flywheel by hand, pulling gently on the starter rope or turning the electric starter on for two or three seconds. Put the spark plug back in place for the winter. This oil treatment will protect internal vital parts over the winter.

4/ *Oil all moving parts.* Run through the check list in drawing number 2/1 and oil every location which should receive oil, E,

P and S all included. Do not lubricate those parts marked to receive grease, however, except if they appear to be dry and unprotected at the moment. Grease tends to become stiff during periods of inactivity. If you put off the greasing until the start of your next lawn mower season, the results will be somewhat superior from a standpoint both of lubrication and ease of operation.

5/ *Dry and then oil all metal parts.* An aerosol can of oil works most easily for this important bit of rust-proofing.

6/ *If there is a storage battery used with your riding mower, have it fully charged.* A full charge acts as anti-freeze protection for it. The newer nickel-cadmium batteries are unaffected by cold weather, although they will tend to lose their charge slowly during long periods of inactivity. The charge will be fully restored after 24 hours on the charger. However, do not leave a battery connected to the charger during a winter storage period. Nickel-cadmium batteries do suffer some damage when charged at temperatures lower than 40°F.

Any sign of corrosion or acid should be cleaned off a storage battery with soap (not detergent) or baking soda solution. Make sure none of the cleaning water drips into the battery's interior.

7/*Block up the tires.* This is particularly important for air-filled tires, but a sound bit of maintenance for any rubber tire which will stand in one place for a long period of time. After all of your other end of season maintenance is completed, gently raise each wheel off the ground by lifting at some strong part of the lawn mower structure. Slip small blocks of wood under the axles or the frame of the mower but *not under the tires,* so the bottoms of the wheels do not touch the ground. On pneumatic tires, inflate them about five pounds over the normal amount of pressure.

SEASONAL
AT THE START OF A NEW SEASON

1/ *Take away the tire blocks.* Inspect each tire first and then make sure that pneumatic tires are restored to their proper pressure before you overlook this seemingly minor detail.

2/ *Sharpen the rotary blade. Or lap the reel cutting edges.*

3/ *Oil and grease all moving parts.* Again consult drawing number 2/1 and your own user manual for guidance.

4/ *Charge the battery, if any.*

5/ *Clean deposits off electrical parts.* This will include both ends of the spark plug, the ignition points and coil wires if they are easily accessible, any starter motor connections which are easy to service, battery contacts, etc. Do not use abrasives such as files or sand paper on ignition points.

6/ *Drain the oil* from your engine's crankshaft and refill with fresh lubricant. This removes the moisture which may have accumulated over the frosty season.

7/ *Fill the fuel tank with clean, fresh gasoline.*

8/ *Start the engine.* This may take more time than normal because fuel has to be pumped or sucked through the entire system. If a fuel line connects separate tank and carburetor units, disconnect it at the carburetor end until fuel begins escaping. An electrical starter powered by a battery should not operate continuously for more than two or three seconds at a single burst. Then allow five or ten seconds for the battery to cool down.

The engine will run rough and smoke a great deal at first. The smoke is due to oil added through the spark plug hole at the end of last season, but that will soon burn away.

9/ *Tune the engine.* Detailed procedures are located in the chapter on troubleshooting, chapter 3, supplemented perhaps by portions of chapter 4 or 5 on engine repair.

CHAPTER 3

Trouble Shooting

TROUBLE shooting is an art based on the axiom that nothing big ever goes wrong. Well, hardly ever. Big lawn mowers and small ones alike are crippled by something small like a loose wire, a faulty spark plug, some loose bolts, a dirty carburetor valve. Or by something rather silly like an empty gas tank, an incorrect control setting, or forgetting how to operate a seldom used piece of equipment properly.

With a minimum of tools but a generous amount of imagination, any well motivated novice should be able to find the source of most of the trouble lawn mowers will give owners in their first several years of operation. Once the source of trouble is pinpointed, you can decide whether to fix the mower yourself or to haul it to the local lawn mower doctor. At least once you've located the trouble, even if you feel that repairing it is beyond your scope at the moment, you won't be a sitting duck for an unscrupulous repairman.

Enlightened troubleshooting technique calls for systematic examination of everything which could possibly cause the trouble which your mower is experiencing. Consistent with the principle that the little things generally go wrong, trouble shooting starts by checking the little things and then works up to the bigger ones. This chapter suggests symptoms to investigate under five major categories of problems which lawn mowers are prone to suffer. The symptoms are arranged very generally from the simple to the difficult. Next to each symptom is the pertinent chapter number to help you correct the problem.

I. ENGINE WON'T START

1. *Ignition switch:* (Chapter 4) Is it turned on or in the proper position for starting?

2. *Stop switch:* (Chapter 4) Is it damaged or inadvertently pushed to where it shorts the ignition line to stop the engine?

3. *Gasoline:* (Chapter 2) Is the fuel tank full? Is it full of gasoline and not some water which you may have put there by accident or some neighbor's kid may have put there as a prank?

4. *Battery:* (Chapter 2) If your mower uses a battery for starting, is the battery powerful enough to turn over the engine? Is it connected and connected properly? Check the user manual for the correct positive and negative battery terminal positions. It is possible that if the connections were reversed, the starter motor might spin but the ignition system would not operate.

5. *Starter:* (Chapter 4 or 5) Regardless of what type of starter is included on your lawn mower engine, check to make sure that it is turning the engine. If the blades or pulleys or gears driven by the engine fail to move when you pull the rope or release the tension on a wind-up starter or engage the starter switch on an electric starter, then the starter itself probably is at fault.

6. *Fuel system:* At this point, unscrew the spark plug and see whether or not the tip is wet with gasoline. If you've been trying to start the engine for several minutes, the odor of raw gasoline should be very obvious. If the plug is wet, the fuel pumping system apparently is working and the problem more than likely is in the ignition system. Problems with carburetor settings could conceivably cause this combination of results; that will be checked further down the list. A dry plug would imply at this point that fuel is not reaching the cylinder, a possibility covered in (I-9).

7. WET SPARK PLUG (ENGINE WON'T START)

7a. *Is high voltage current reaching the spark plug?* Remove the spark plug wire from the spark plug and hold the metal contact about 1/8″ away from some exposed metal part of the engine block. Spin the flywheel by hand or gently pull on the starter rope or spin the electric starter for two or three seconds. A bright, flame-like spark should jump the 1/8″ gap and cause a snapping sound. If it does, check (I-8). If there is no spark, the ignition system is faulty somewhere along the line. Look below for:

7—WET SPARK PLUG BUT NO SPARK.
8—WET SPARK PLUG AND SPARK.
9—DRY SPARK PLUG.

7b. *Spark plug:* (Chapter 4 & 5) Are the electrodes relatively clean and unpitted? They should not appear crumbly. You can clean the old plug with a wire brush, then make sure the electrodes are set the proper distance apart (.030″ for Briggs & Stratton, .025″ for Clinton and *usually* for Kohler). A new plug costs so little, however, and replacing the spark plug will virtually eliminate this as a possible source of trouble.

7c. *Armature gap:* (Chapter 4) If the bolts holding the magneto armature in place loosen up, the armature assembly could slip too far away from the magnets within the flywheel and thus be unable to generate a spark. Loose bearings on the flywheel also could cause this trouble in an old or abused engine. To reset the armature gap, follow the instructions in Chapter 4. To test for worn bearings at the flywheel end of the crankshaft, set the armature gap while shoving the flywheel firmly toward the armature. If the gap fluctuates by more than .010″ when you shove in the opposite direction, you should suspect a worn bearing.

7d. *Flywheel key.* (Chapter 4) The metal key which helps hold the flywheel into its precise location for proper magneto performance can be sheared off or bent by rough handling, engine malfunction, mowing equipment malfunction, excessive vibration, or simply by old age. With the key broken or worn, the flywheel often shifts a bit to the right or left of its proper position and the permanent magnets in the flywheel will not pass the magneto armature at the right time.

7e. *Ignition points.* (Chapter 4) Ignition points seldom become defective or slip badly out of position suddenly. If you were experiencing starting difficulties before or rough and inefficient engine performance the last several times the mower was used, ignition points or condensor may very well be the problem. The points on large engines are easy to reach but on small Techumseh, Lawn Boy, Clinton or Briggs & Stratton engines, you must pull off the flywheel to reach them. They can be cleaned with a fine file or fine emory cloth, but this is only a very temporary solution. If points are so dirty or pitted that they require abrasive cleaning, they should be replaced at your earliest opportunity.

7f. *Condensor.* (Chapter 4) Like the points, the condensor seldom becomes inoperative suddenly. Since the points and the condensor function as a team and are mounted in close proximity to each other, when one is replaced the other generally is too. A sudden ignition failure could be caused by loose or broken wires on the condensor or points.

8—WET SPARK PLUG AND SPARK (ENGINE WON'T START)

8a. *Flooding.* An excess of gasoline can at times be drawn into the cylinder and inhibit starting even though all of your engine's mechanical and electrical components are in good enough condition to start the engine. If you detect an obvious odor of unburned gasoline after several unsuccessful attempts at starting an engine, let the lawn mower alone for a few minutes so the excess gasoline will evaporate.

8b. *Water in fuel.* (Chapter 2) Since gasoline floats on top of water, before the fuel pump could suck up gasoline, it would first draw up any water in the tank. Therefore you must check the *bottom* of the tank for signs of water. One way to accomplish this is to insert a soda pop straw to the bottom of the fuel tank and withdraw a small sample; if it doesn't smell like gasoline or if it separates into two distinct although colorless layers, water probably is in your fuel system. Once the fuel tank has been purged of the water, there may still be water in the fuel line. By working the starter until the smell of gasoline is detected at the open spark plug hole, you will know that the water has been pumped out and the gasoline pumped in.

8c. *Air cleaner.* (Chapter 2) Gasoline is but one ingredient needed for combustion. Air is the other. If the air cleaner is exceptionally dirty, it can retard the flow of air enough to prevent starting. Despite the practice prevalent among many mechanics, engine manufacturers recommend that you *do not run an engine without the air cleaner for even a few seconds.* Dirt can be sucked directly into the cylinders and cause very serious damage.

8d. *Over-choking.* (Chapter 4) Just as a blocked air cleaner can prevent an adequate supply of air from reaching the combustion chamber, a choke can cause the same effect. If your engine receives too little air through the carburetor, the condition is known as over-choking.

On automatic choke models, remove the air cleaner to make sure that the choke disc plus connected rods and springs can be moved freely with your finger or by some tool which can not possibly slip into the carburetor throat accidentally. With the air cleaner back in place, hold the automatic choke almost closed. On some models it may be necessary to leave a tool such as a screw driver in the carburetor to hold the choke disc open after you replace the air cleaner. (This is for *test* purposes only.) Then gradually let it open more and more with each successive starting attempt. If this manual over-ride works, you probably will have to disassemble the carburetor or choke according to the methods described in Chapter 4.

Choke-a-matic or manually choked models can be manipulated on the mower handle, on the dust cover or on the engine itself. Although you normally place the control in "START" or "CHOKE" when starting the engine, move it instead to "FAST" or "SLOW" and try starting. On warm days in particular, the choke is not generally needed after the first several attempts at starting the engine. The choke is almost never needed to start the engine when it is warm, such as after pausing to refill the fuel tank or to catch some refreshments.

8e. *Carburetor mixture is too rich?* (Chapter 4) If *too much fuel* is being sent through the carburetor, the net effect can be practically identical to *too little air,* as in the case of a blocked air cleaner or over-choking. Locate the needle valve control and tighten it clockwise to its fully closed position. Remember the approximate number of turns it originally had been set to. Then reopen the needle valve 1-1/2 turns. After two or three attempts the engine could at least sputter and feel as though it were ready to start; if so, open the needle valve 1/2 turn more and work the starter again. Continue this until the needle valve is back to its original position. If the engine starts midway through this procedure, the carburetor must be set more precisely according to the procedures in Chapter 4.

8f. *Carburetor valves stuck open?* (Chapter 4) The carburetor must be taken apart to make this determination.

9—DRY SPARK PLUG (ENGINE WON'T START)

9a. *Fuel line blocked or loose?* (Chapter 4 or 5) Disconnect the fuel line at various points to determine how far gasoline is able to flow freely or to be pumped when the starter is activated. Also check for loose connections, bent or pinched tubing and dirty or shut-off valves inadvertently closed.

9b. *Carburetor setting.* (Chapter 4) Locate the needle valve control on the carburetor and open it an additional 1/2 turn. Try the starter two or three more times and then open it yet another 1/2 turn. Continue this process for two complete turns. If no success is encountered, return the needle valve control to approximately its original position.

9c. *Carburetor valves.* (Chapter 3 and/or 4) If check valves or fuel pump valves or carburetor valves are stuck shut, they could block off gasoline. The carburetor has to be broken down for a check-up and cleaned if necessary.

9d. *Fuel pump.* (Chapter 4) If the fuel pump is mounted directly on top of the gasoline tank, as is the case with small Briggs & Stratton engines, when the carburetor valves are checked, the fuel pump will also be checked. On engine models with a separate fuel pump, it should be investigated immediately after the carburetor.

II. ENGINE IS HARD TO START

Most of the preceding conditions which could cause an engine not to start, might also make it *difficult* to start. There are several situations beyond the scope of the engine itself, however, which could make starting quite difficult.

1. *Equipment drag.* (Chapter 6, 7 or 8) If power take-off equipment or self-propulsion equipment is engaged when you attempt to start an engine, it will make the endeavor extremely difficult. Not only must the starter spin the *engine* fast enough to initiate combustion and ignition, you will also be using the starter to power whatever *equipment* is engaged.

2. *Loose blade.* (Chapter 6 or 8) (Caution: Disconnect spark plug wire before checking for a loose rotary mower blade.) Not only will starting be difficult with a loose belt or, on a rotary mower with a loose blade, it could prove to be dangerous as well. A loose blade can generate enough momentum during a backlash to counteract the flywheel, causing a kick-back—the engine and starter will momentarily move backwards. With the flywheel held firmly in place, if the blade can move at all, the blade is attached too loosely.

3. *Clutch or transmission.* (Chapter 8) If the clutch or transmission is engaged or dragging, it will cause drag on the engine, making it difficult if not impossible to start.

III. MOWER ENGINE MISSES OR PERFORMANCE IS ROUGH

Some of the conditions already mentioned in the fuel and ignition check points, if not severe, might cause the engine to put-put-put or threaten to stall or idle as rough as an old tractor. However, don't rule out the possibility that equipment connected to your engine can be malfunctioning.

1. *Fuel tank almost empty.* (Chapter 2) When the tank is down to its last few drops on certain carburetion systems, rough

turf, slopes, turning or some similar condition can cause the fuel to flow erratically. The correction is an obvious one—fill the tank. Running out of gas can lead to more than frayed tempers. Sediment, gum, varnish, droplets of water and other foreign matter which gets into a fuel tank or which forms inside the tank, usually stay harmlessly out of the way. If they are sucked out of an almost empty tank and into the fuel system, they plug needle valves, gum diaphragms and create a general nuisance.

2. *Ignition shorted?* Those fancy covers and trim which grace the outside of so many lawn mowers of all sizes and descriptions, often get fatigued or bent or improperly closed. Ignition wires—spark plug, coil, magneto, battery—often are attached too close to the metal shrouds and hoods. If rough performance is plaguing you, check that none of the ignition wires can brush against a metal part.

3. *Spark plug.* (Chapter 4 or 5) A dirty or oily spark plug, or one with the wrong gap setting, can create rough engine performance.

4. *Choke.* (Chapter 4) If a control cable, spring or connecting rod on the choke is loose, the choke could be bouncing open and shut as you bounce across the wide green turf. The result could be a bouncy sound to your engine.

5. *Carburetor.* (Chapter 4) If the mixture is too lean, the engine may stall momentarily or run very rough whenever the load increases. *You* might not notice an increased load when the *engine* does. Move the carburetor's needle valve control out (counter-clockwise) 1/2 turn and see if the performance improves. If it improves, but not enough, move the control another 1/2 turn. Continue until the engine handles well.

6. *Tight mower reel blades* (Chapter 7) If just one blade out of a five or six blade reel mower is badly out of alignment, it could create a jerking action which you might mistake for rough engine performance.

7. *Loose or broken equipment.* (Chapter 8) If slow moving chain drives and sprockets are out of line, or slow moving gears don't mesh properly, or if slow moving transmission or differ-

ential parts are broken or loose, the resulting jerking sensation might be mistaken for a missing engine. Frayed, worn or stretched belts in the power train, and even worn or flatened tires, can create an erratic performance.

IV. VIBRATION

A certain amount of vibration is to be expected whenever moving parts are involved. But when the vibration becomes noisy, uncomfortable or threatening, then you'd better decide that something is wrong and try to find out what.

1. *Cutter blade.* (Chapter 6) On a rotary mower, the blade is the first and most logical place to look when vibration strikes. A bent blade rotating at approximately 3000 r.p.m. can kick up quite a jiggle. A blade which has been badly knicked by a stone or which has been sharpened but not properly balanced will kick up even more troublesome vibration. Running a mower with a serious vibration problem can only lead to even more serious difficulties such as a broken blade adapter, bent crankshaft, prematurely worn bearings. There is no sure way of telling which bolts will be shaken loose by the vibration, including the very bolts which hold the blade to the crankshaft.

2. *Loose bolts.* (Chapter 6, 7 or 8) The number of locations where a bolt might be loose is breath taking. But the ones most directly connected to the power should be checked first. For example:

- The engine mount
- Blade mount
- Power take-off attachments
- Reel cutting blades
- Drive wheels
- Wheels and axles not directly connected to the engine
- Transmission, transaxle, differential
- Flywheel

3. *Cracked deck.* (Chapter 6 or 8) The decks on rotary lawn mowers sometimes can be split or cracked by twisting the frame,

vibration, rocks being thrown against them and in sundry other ways. Not always with justification, lawn mower assemblers would consider each of these situations negligence on your part so you wouldn't stand much of a chance to get satisfaction from the company. A cracked or split deck could create a very bad vibration and a potentially dangerous condition.

4. *Wrong blade or blade adapter.* (Chapter 6) Lawn mower manufacturers change the design of blades and blade adapters so often, there is a fairly good chance, therefore, that you might slip on a blade not made to fit the particular adapter on your machine. This is especially likely to happen on private label mowers which department store chains often sell under their own name but often purchase from several different prime suppliers. The vibration caused by an incorrect blade or blade adapter might be slight. On the other hand, it could be frightening.

5. *Bent crankshaft?* (Chapter 4) This is a rough one. It takes just a slight bend for a crankshaft to kick up a lot of extra vibration, particularly on a rotary mower. If you've checked all other possible sources for bad vibration, and if you suspect the crankshaft is at fault, two limited choices are open to you. One, you can take the machine, or at least the engine, to a competent and well-equipped mechanic who can check it for you with a gauge designed for that very purpose. On the other hand, you might just replace the crankshaft with a new one; this can be an expensive alternative unless you are certain that a bent crankshaft is at fault.

V. LOSS OF POWER

Psychological factors play a big role when a lawn mower owner decides that Old Betsy is feeling a bit down on her oats. Maybe the neighbor got a shiny new model for his birthday and suddenly your own two year old mower feels inadequate. Even how *you* feel on a particular day affects how you think the mower is functioning. Consequently, loss of power is one area in which it is wise to put off any rash action brought on by a feeling that the old mower is slipping in the power department.

1. *Changing conditions.* Moist days can make a mower engine perform with more power than dry days. A hot engine, within certain bounds, performs with greater efficiency than a cool engine. At certain times of the year your grass is heavier or harder to cut than during the rest of the growing season. If your lawn has grown more than usual between cuttings, the engine will have to work harder than usual.

2. *Subtle obstructions.* (Chapter 2) A build-up of grass or other debris on any moving parts such as the blades, gears, chains, sprockets or on the inside portion of wheels will waste your mower's power. And on rotaries, an accumulation of grass under the hood can directly or indirectly slow down the blade or force the engine to work harder.

3. *Worn parts.* (Chapter 6, 7 or 8) On self-propelled mowers, tires and drive wheels may gradually wear out; they then can spin either noticeably or subtly and rob your mower of significant useable power. Belts can wear thin, stretch or slip out of adjustment and create a loss of power.

4. *Lubrication.* (Chapter 2) Even if you have faithfully followed the routine maintenance recommended by the lawn mower assembler and by this book, there is still a chance that some grease cup or some bearing needing oil has been generally overlooked. With the spark plug wire anchored safely out of the way, test all of the various moving parts. If you find an especially stiff one which shouldn't be stiff, find a way to lubricate it. It may first be necessary to apply a solvent such as "Liquid Wrench" to loosen the bearing or to soak out dried oil and grease.

5. *Equipment drag.* (Chapter 6, 7 or 8) If you've added more accessories or equipment since your mower was brand new, of course it will appear to have less power. More of its power will be spent in twisting or turning the new gears and grinders and gadgets. You might also have adjusted your reel blades in an effort to make them self sharpening but tightened them too much. The net result can over work an engine barely able to handle

the load anyway. Even a grass catcher or leaf catcher can add drag to a mower as it fills up.

6. *Governor.* (Chapter 4) Oil, dirt, grease, grass and other foreign matter lodged on an air vane governor could keep the spring from pulling it fully open. Thus the engine would never be able to operate at full power. General wear and tear also have to be checked periodically with either the mechanical or air vane governors.

7. *Poor compression.* (Chapter 4) Chapter 4 tells how to test for faulty compression and how to cure many of the ailments for which this is a symptom. Listed in the order they most frequently occur, they are:

- Loose spark plug or damaged spark plug gasket
- Loose cylinder head bolts
- Blown or damaged head gasket
- Burnt valves or valve seats
- Insufficient tappet clearance
- Over-sized cylinder or worn piston rings
- Warped engine head or valve stems.

CHAPTER 4

Engine Repair: The Briggs & Stratton Engine

THE mysterious, greasy, noisy world beneath that painted cover on your lawn mower engine need not intimidate you. Driving a car, shaving and baking cakes were mysterious, too, until you learned how. Repairing or overhauling, the engine on a lawn mower is principally a matter of having the proper instructions available along with a modest selection of the proper tools. Fortunately, both items are inexpensive and readily available in almost every part of the country.

Approximately two-thirds of all the different makes of lawn mowers sold today are powered by a Briggs & Stratton engine, maybe more. Briggs & Stratton has a reputation for making quality small gasoline engines in Milwaukee almost from the day brewers started making beer out there. Somewhat akin to the image Rolls Royce used to hold about itself, Briggs & Stratton is very self-conscious about maintaining a shiny reputation for quality and service. It sends qualified mechanics to its own school for three weeks to study Briggs & Stratton engines. And it keeps up-to-date a massive repair and maintenance book covering virtually all aspects of its many engines. The company even makes classroom instruction material available for high school use. Unlike some engine manufacturers, Briggs & Stratton will sell the average Briggs user a copy of the repair manual for a very nominal cost.

In addition to a network of several hundred local repair shops authorized by Briggs & Stratton to tackle repair and tune-up work, the company also maintains 42 centralized service and parts distributors in 30 of the States and three Provinces of Canada. Each central distributor stocks an inventory of both engines and parts, as well as employing a staff qualified to tackle jobs more difficult than many local mechanics care to handle.

Simply because Briggs has had quite a sterling reputation doesn't guarantee that it will always deserve acclaim. In fact, there are a few indications that Briggs & Stratton is giving in at least a bit to the industry's demand for quantity with some sacrifice of quality.

The Kohler Corporation seems to supply the power package on many of the larger mowers and home tractors not packing a Briggs & Stratton engine. Clinton and Tecumseh-Lauson also supply a significant number of both large and small engines.

MODEL NUMBERS ON BRIGGS & STRATTON ENGINES CAN HELP you determine what features the engine includes and can even provide you with a clue as to whether or not some apparently nameless engine is or is not a Briggs & Stratton creation. Some lawn mower assemblers take a Briggs & Stratton engine—and in some cases, another manufacturer's engine—and paste on a new label or bolt on a fancy cover so they appear to be manufacturing their own engine. There are five basic elements to the Briggs & Stratton model number.

SAMPLE CODE	**9**	**2**	**9**	**0**	**5**
CODE POSITION	**1**	**2**	**3**	**4**	**5**

In the sample shown above, a hypothetical engine with model number 92905, code position one refers to the cubic inch displacement of the engine. In this example, the displacement is 9 cubic inches, which converts roughly to 3.5 h.p. This first code position, and only the first, can in fact be two digits. An 8 h.p. Briggs engine, for example, has a 14 cubic inch displacement; a rig of that size might be given a model number of 142905 if the rest of its essential parts were similar to its smaller brother used here as an example.

"Digit 1" Cubic Inch Displacement & Horsepower	Digit 2 Basic Design Series	Digit 3 Crankshaft, Carburetor, Governor	Digit 4 Bearings, Reduction Gears & Auxiliary Drives	Digit 5 Type of Starter
6 2 hp	0	0–	0– Plain Bearing	0– Without Starter
8 3 hp	1	1– Horizontal Vacu-Jet Air Vane	1– Flange Mounting Plain Bearing	1– Rope Starter
9 3.5 hp	2	2– Horizontal Pulsa-Jet Air Vane	2– Ball Bearing	2– Rewind Starter
10 4 hp	3	3– Horizontal Flo-Jet Pneumatic	3– Flange Mounting Ball Bearing	3– Electric: 110 volt Gear Drive
13 5 hp	4	4– Horizontal Flo-Jet Mechanical	4–	4– Electric Starter-Generator: 12 volt Belt Drive
14 6 hp	5	5– Vertical Vacu-Jet Air Vane	5– Gear Reduction (6 to 1)	5– Electric Starter Only: 12 volts Gear Drive
17 7 hp	6	6–	6– Gear Reduction (6 to 1) Reverse Rotation	6– Windup Starter
19 8 hp	7	7– Vertical Flo-Jet Air Vane	7–	7– Electric Starter & Alternator: 12 volt Gear Drive
20 8 hp	8	8–	8– Auxiliary Drive Perpendicular to Crankshaft	8– Vertical-Pull Rewind Starter
23 9 hp 24 10 hp 30 12 hp 32 15 hp	9	9– Vertical Pulsa-Jet Air Vane	9– Auxiliary Drive Parallel to Crankshaft	9–

Illustration 4/1. A 5- or 6-digit number, such as 192902, breaks down into 5 bits of information thus: 19/2/9/0/2. The "19" indicates an 8 h.p. engine, the "2" represents a basic design series which is of little consequence to the consumer, the "9" identifies that the sample engine employs a Pulsa-Jet carburetor and an air vane governor, the "0" tells the type of bearings used, in this case a simple one, and the final "2" tells that the engine was outfitted with a rewind starter when it left the factory.

Number two code, "2" in our example, identifies basic design features such as ignition systems, cylinder construction and other esoteric information of concern to Briggs but of little immediate importance to lawn mower users. The most popular small lawn mower engine, the 3.5 h.p. model, generally has a "2" in this second position.

Third down the line is a code which identifies the carburetor, governor and direction of the crankshaft. Reel mowers often require a horizontal crankshaft, code numbers 1 through 4. And rotary mowers with a vertical engine are assigned codes 5 through 9.

Code position four talks about bearings and related features. In most cases this will be "0" which stands for a plain bearing. Tractor type equipment, as well as some self-propelled mowers, could have an "8" or a "9" in this slot to show that they have an auxiliary drive shaft. This is available for power take-off (PTO) to blow snow, rake leaves or even to drive the wheels if the main crankshaft is coupled to a rotary mower blade. In that case, "8" shows that the auxiliary drive is perpendicular to the crankshaft and "9" that it is parallel to the crankshaft.

Starters are identified by the final code digit and they vary immensely according to taste preferences and cost. Illustration 4/1 categorizes the code numbers for starters and all other coded features.

Sears Roebuck & Company, according to the Briggs manual, uses Briggs & Stratton for power on some of its items but adds a "500" in front of the Briggs model number. Sears model 500.20054, for example, would convert to a Briggs & Stratton model 20054. A good many manufacturers simply adopt the Briggs engine model number as their own part number. *Yardman,* for instance, on its Model 2320-2 20-inch rotary mower lists part number "92905, engine." Now that you're privy to the code, you can read that part number as a 9 cubic inch (3.5 h.p.) Briggs engine with a vertical crankshaft and Pulsa-Jet carburetor (code 9), a plain bearing (code 0), plus a 12 volt electric starter (code 5).

Sensation, on its Model 21G6 21-inch mower, gives 92902 as the part number for its engine. Consulting the Briggs model code chart, you can quickly ascertain that it is the same engine as Yardman mounted on their 20-inch rotary mower except that the starter code for Sensation's version is "2" instead of "5", which means a rewind hand-powered starter instead of the more deluxe battery-powered starter.

Allis-Chalmers has a model WB1930 mower which lists part number 92502 for its engine, which is, in fact, a Briggs & Stratton 3.5 h.p. engine, very similar to the one mounted by Yardman and Sensation in the above two examples. In the carburetor code position, however, there is a "5" showing that Allis-Chalmers chose an engine which uses a Vacu-Jet carburetor instead of the Pulsa-Jet type used on the engines for Yardman and Sensation machines mentioned above. This is not an especially significant variation.

THE LARGE LAWN MOWER ASSEMBLERS buy so many engines from Briggs & Stratton (or any other engine manufacturer) that they can obtain special construction features. Sometimes the special feature may relate to a technical detail or two which they've discovered will result in smoother operation of your lawn mower. Often, however, the special feature is selected by the lawn mower manufacturer to make his assembly operation easier or his price lower. Although these special engines are described in the Briggs & Stratton service manuals, they generally are not stocked by regional parts and service centers. Consequently, replacement motors and parts might have to be shipped from Milwaukee whereas a stock engine and its replacement parts might very well be sitting on the shelf in 39 different parts of North America.

According to the Briggs service manual, *Jacobsen* employs a special version of a 92908 engine on at least one of its lawn mowers. *Toro Manufacturing Company* uses two different variations of the basic models of the 92908 engine on its Fiesta lawn mower, according to the Briggs manual; each of the versions uses a different, special-order crankshaft. The Briggs manual also lists *Hahn-Eclipse* and *Simplicity Manufacturing Company* on its page of special engines. Hahn-Eclipse orders its own version of a 92988 engine and Simplicity a 92998.

Additional special order engines for small tractors put out by several concerns are listed in the Briggs manual: *Jacobsen's* Tractor A5311 and Tractor A53135; *Ford Motor Company's* Tractor 70, Tractor 75 and Tractor 80; *Simplicity's* Tractors model 717, 727, and 728; *Allis Chalmers'* Tractor B-207.

Since the 3.5 h.p. Briggs engine costs about one half the total price of a complete lawn mower, you won't want to replace the engine often. If replacement does become necessary, it will probably happen exactly when you most want the equipment to be operating. A new crankshaft for a 3.5 h.p. engine generally is priced quite reasonably. Installation is extra, of course. The special crankshafts for the special engines described in the above paragraphs require special orders. The Briggs service manual includes a coded stockroom guide next to almost every part it lists. Most of the special crankshafts listed are given a rating of 4 on a scale which ranges from 1 to 5. Stockroom clerks are advised by Briggs that "4" means "slow moving part. To be stocked if warranted by local demand." A few of the special crankshafts are not to be found in some lists at all, which means your local supplier then must send in his order and wait for the factory to ship one or, worse still, make one.

Waiting several weeks for a new crankshaft because it is a special order part is enough to make you want to go out and buy a new mower, isn't it? No doubt some lawn mower assemblers have designed their equipment with this in mind.

A BIT OF BASIC THEORY ABOUT GASOLINE ENGINES should make matters simpler before getting into actual repair procedures. Most lawn mower engines are called 4-stroke cycle engines, which generally gets shortened to just "4-cycle." That means merely that the engine goes through four separate strokes to get one stroke of power. The 4-cycle engine is not as inefficient as it may seem. Detroit's automotive geniuses have lived with the concept for years. The 4-cycle engine which powers your lawn mower works on the same basic 4-cycle theory as the one which over-powers your car.

At the heart of a gasoline engine is the piston which slides up and down inside a tight fitting cylinder. One end of the piston is connected to a crankshaft which spins around to power

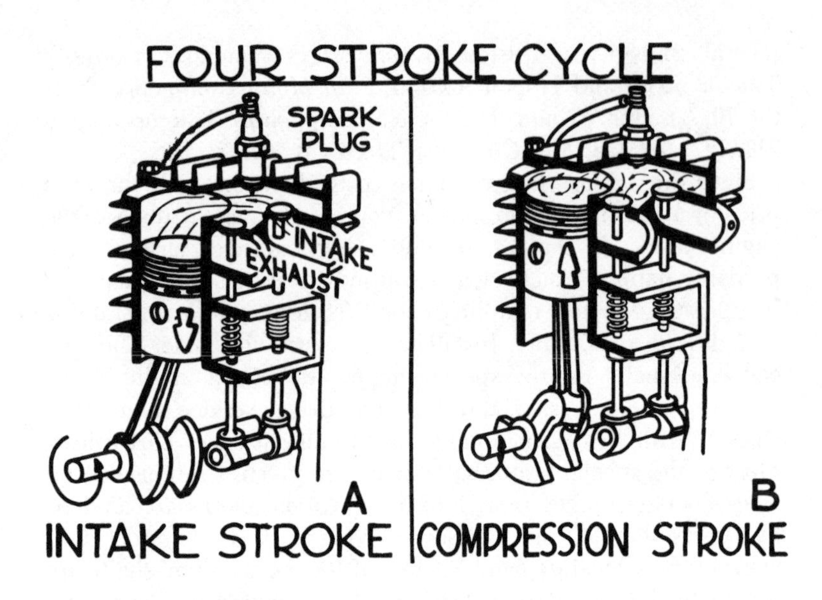

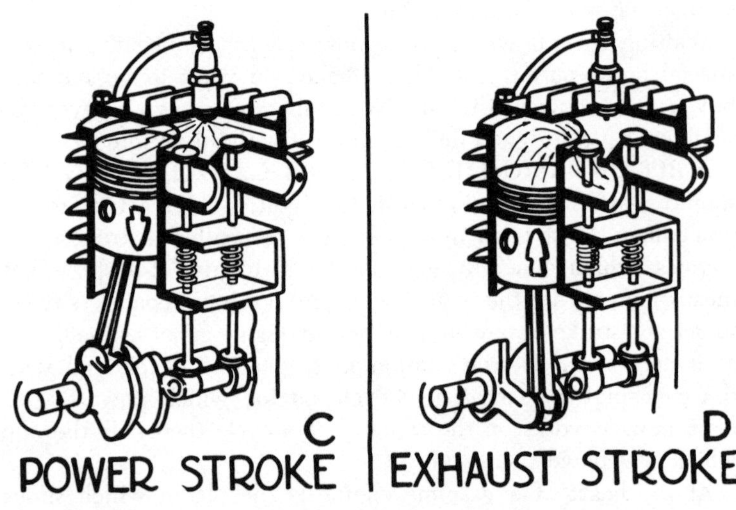

Illustration 4/2. Theory of Operation for a Four-Stroke Cycle Engine. During the INTAKE stroke, a mixture of gasoline and air is drawn into the cylinder because of a vacuum generated by the piston's down-stroke. The gasoline-air mixture is next compressed, in the COMPRESSION stroke, by the piston's up-stroke. During the POWER stroke the fuel mixture burns because the spark plug has fired; the piston is forced down by the expanding combustion products. And in the EXHAUST stroke, the piston's up-stroke forces out the burned gases through the exhaust valve. The INTAKE stroke follows immediately and so forth.

your car's wheels or your lawn mower's blades. Your car's engine has four, six or eight pistons while your small lawn mower engine has but one.

Pistons in your 3.5 h.p. lawn mower engine are about 2-1/2 inches in diameter and are machined to within 1/1000th of an inch of the specified size, so accurately must the piston match the cylinder. Drawing number 4/2 will help you visualize the mechanical layout of the various parts involved in a gasoline engine.

On the *intake stroke,* while the piston is pulled down, a mixture of air and gasoline is sucked into the cylinder through the *intake valve* which is open only during this stroke. After the intake valve closes, the piston is forced upwards into the cylinder again, compressing the air and gasoline; appropriately enough, this stroke is called the *compression stroke.*

The *spark plug,* which sits atop the cylinder in our drawing, is fed a charge of high voltage electricity at the proper instant and the resulting spark ignites the gasoline, beginning the *power stroke.* Technically this is not an explosion, but the mixture of air and gasoline does burn mighty fast and the result is a rush of hot gasses which pushes the piston downwards once more. The force which can be generated within a single cylinder only 2-1/2 inches in diameter is enough to keep the engine going through the three powerless strokes and at the same time spin your rotary lawn mower blade almost 3000 r.p.m.

Still coasting from the power stroke, the piston charges upward once more and at the same time, the *exhaust valve* opens up. The burned up gasses are forced out, the exhaust valve then closes, and our engine has successfully completed its fourth stroke, the *exhaust stroke.* Then it begins all over again on the intake stroke. On an engine turning at 3000 r.p.m., this progression of four strokes is repeated 25 times every second!

A steel shaft called the crankshaft receives the up and down thrust of the piston and converts it into a spinning motion. But a piston jumping straight up and down, no matter how fast it might move, could never produce a smooth rotating type motion. To add smoothness to the engine's frenetic pace, a relatively heavy flywheel is fastened to one end of the crankshaft. The flywheel is set spinning by each power stroke and soon generates enough inertia through its weight to keep the crankshaft, valves and pistons moving smoothly during the three unproductive strokes.

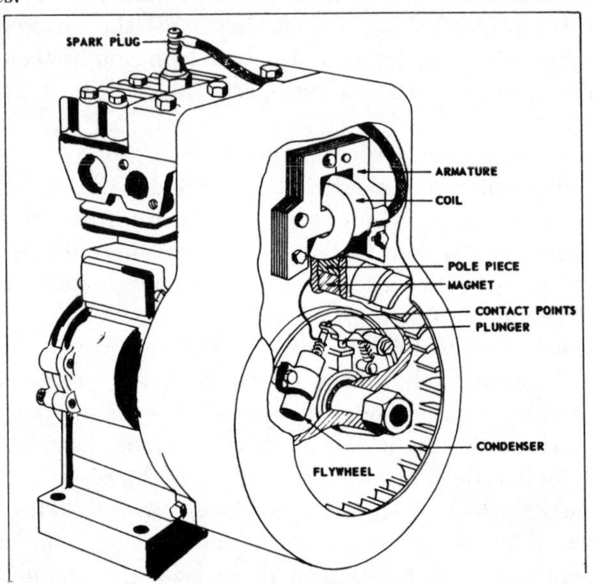

Illustration 4/3. The Flywheel Ignition System. All of the essential parts for producing high-voltage current for the spark plug are built into the flywheel or are located immediately beneath it on most small gasoline engines.

The flywheel on small engines has been designed to do much more than simply spin its weight around. Fan blades are cut into the flywheel's outside surface so it can double as a blower. The burning of gasoline inside the cylinders generates a great deal of excess heat and the blower part of your flywheel helps to cool the aluminum piston and the aluminum block which forms the cylinder on most small engines. In your car, with the exception of small foreign built models, water circulates in the block to cool the metal parts. The water, in turn, is cooled in the radiator. Small gasoline engines, having but one cylinder to cool, rely very nicely upon the air to absorb excess heat.

Briggs & Stratton, at least on its small engines, gives the flywheel still one more chore to perform—it acts as one half of a magneto. A small but very powerful permanent magnet is built into the inside of the flywheel. As the magnet moves past a coil of wire, it builds up an electric charge in the coil very much like a generator would. The basic difference between a *magneto* and a *generator* is that the magneto employs a permanent metal magnet while the generator uses an electro-magnet to build up electric charges in a coil of wire.

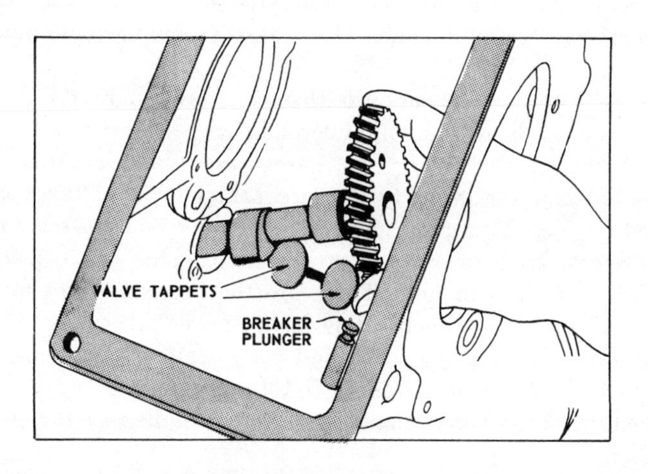

Illustration 4/4. Camshaft Gear. Attached to the drive shaft, the camshaft "switches on" the intake valve, exhaust valve and spark plug.

The magneto's charge is used to energize the spark plug at the precise moment the piston is ready to begin its next power stroke. A very sensitive switch is necessary to supply the electric charge at precisely the proper moment, about 25 times every second. If it comes too late or too early, the engine's performance will be rough and efficiency reduced. The switch must at the same time be very rugged since the amount of current it must carry can approach 10,000 volts. The intake valve and exhaust valve likewise must open and close at precise intervals which coincide with the piston's up and down movements.

A gear attached to the crankshaft drives a set of cams at one half of the crankshaft's speed. One cam opens the intake valve at the appropriate instant once every two turns of the engine. Another cam does likewise for the exhaust valve. A third cam operates a plunger which for but an instant indirectly switches on the current which feeds the spark plug.

LAWN MOWER MANUFACTURERS DO NOT HAVE THE AVERAGE HOME HANDY OR UNHANDY MAN IN MIND when they design lawn mowers. In fact, most don't seem to have any repairman in mind. What interests them almost exclusively is their own assembly line. That's repair complication number one.

Number two complication is that your lawn mower has more than likely evolved through numerous changes (forward or backward) over a number of years. Parts were added, altered or removed. Some variations in size are called for, but there is no sound engineering reason why any assembler really needs to put 17 different kinds of cap screws into its machine not to mention 11 different kinds of nuts. Those are the precise figures for the G. W. Davis Inc. 32 inch riding lawn mower taken right out of their own "Owner's Operation and Care Manual." Nuts and cap screws include 1/4th (4/16ths), 5/16ths and 3/8ths (6/16ths) of an inch. And the Davis manual was simply pulled from a pile at random.

The manual for a Yardman Model 3400-2 riding mower, also picked at random, shows that at one time somebody evidently made a studied effort at standardizing parts. The 5/16ths inch nut, for instance, is used almost everywhere—but on places where the switch would have no engineering purpose at all, suddenly a 1/4th or a 3/8ths inch nut appears in the technical drawings.

Part of the explanation for lack of standardization within individual lawn mowers is just lack of any obvious concern for repair problems. The other part of the explanation is that nobody *manufactures* lawn mowers. They're *assembled* from motors and transmissions and differentials and wheels and cables and even handles or blades made by a myriad of different fabricators. If the little old handle maker has always used a 3/8ths inch drill, and the little old engine maker has always used a 5/16ths inch drill and the little old wheel maker has always used a 1/4th inch drill, finally the lawn mower maker will have to use bolts of three different but almost identical sizes.

When the typical unschooled handyman heads for a lawn mower which has dozens of different sized bolts and nuts, he's likely to pick up an adjustable wrench, one of thost do-it-alls which can open up to one inch wide or down to 1/16th inch narrow . . . well almost. They never quite hit the exact size on the head. After loosening or tightening any bolt several times with an adjustable wrench, the once square edges are rounded off to the point where the adjustable wrench no longer will catch on it.

It pays you to invest in some tools at the same time you invest in a lawn mower. If the lawn mower assemblers would take the trouble to standardize, they could very well afford to include two or three wrenches of the proper size with each mower. They could even stencil their name and logo on each wrench. Since they don't, you'll have to supply yourself with a set of tools.

Illustration number 4/5 lists three different collections of tools. The first list includes tools which any lawn mower owner ought to have just to take care of routine and seasonal maintenance. List number two should be of interest to the person who likes to save money, or to the man or woman who gets a kick out of tinkering, or to the person who doesn't have a good repair station within easy reach. The third list of tools is foı the man or woman who wants to know how and why everything works and insists on doing it himself. The professional lawn care person who puts in a lot of time behind a mower falls into this third category. He must often do his own repairs because he can't afford down-time caused by an ailing mower in the shop.

Repair instructions which follow also fall into the same three very broad categories as the tool lists. It is important that you not be frightened away by the aura of mystery and engineering expertise and space age technology talk which surrounds the lawn mower. It's built no better than some of your kids' science toys!

Illustration 4/5

A List of Tools for Lawn Mower Repair

A/ *The Elementary Equipment*
which any owner of lawn mowers should possess.

spark plug wrench
screw drivers—small, large and Phillips
pliers
oil—can of light oil and aerosol can
pocket knife
spark plug gauge
hammer
a small set of box wrenches or open-end wrenches
grease—small tube

B/ *For the Handy Person*
who is ready to try more advanced repair techniques.

drills (and a 1/4-inch electric drill)
screw drivers—a larger selection of regular and Phillips.
torque wrench
flywheel puller or large gear puller
feeler gauge
steel ruler—accurate to 1/32nd inch
vice grip pliers
piston ring compressor
files—coarse and fine, pointed and flat
wire brush
needle nosed pliers
. . . . in addition to all of List A

C/ *Advanced Equipment*
which is needed for sophisticated engine repairing

starter clutch wrench (on engines which require one)
flywheel holder
tachometer
taps—set of metal cutting threaders
punches—set of thin metal punches
vice
valve spring compressor
valve lapper
chisel—small metal chisel
micrometer
volt-ohmeter
wrenches—assortment of open end and box wrenches,
or set of socket wrenches (the latter is preferred)
. . . . in addition to all of List A and List B

If you drive a car, operate an Instamatic camera or run a sewing machine, you can keep the family mower properly maintained. On a higher level of difficulty, if you can bake a good angel food cake, operate a Polaroid camera or understand your sixth grader's science book, you can make almost any repair your lawn mower will encounter in its early years and successfully diagnose the more complicated repair problems which require specialized or expensive equipment.

Approach lawn mower repair gradually and systematically. Read this book and accumulate the basic tools. Move progressively into harder and harder parts of lawn mower repair, and as you hit the more advanced levels, you'll probably want to invest in the repair and parts manual for whichever engine is mounted on your own mower. Unless you already have well developed fix-it instincts, you'd better find out what you're doing before doing it. Repair shop after repair shop reports that the number one cause of engine breakdowns is *improper use.* And the second most frequent complaint is *accidental misadjustment;* somebody fiddled with a control or a screw or an external mechanism he didn't understand and made matters worse instead of better.

THERE ARE SEVERAL BASIC CARBURETORS, ignition systems, governors and lubricating mechanisms found on Briggs and Stratton engines. It is important to identify the type of equipment featured on your particular mower. The Briggs model number can be of help. Exploded drawings on near by pages will also prove valuable. It is amazing to note how very similar the various engines shown here are regardless of whether it is a 3.5 h.p. vertical crankshaft model or an 8 h.p. vertical crankshaft model. Only the size of various parts is different in most cases.

When ordering replacement parts, assuming you have not purchased a current Briggs & Stratton parts and service manual, you may specify the reference number shown in these exploded drawings and parts list plus the model and type number of your engine. In addition to the model number, such as 92902, *Briggs also requires a type number,* such as 0061 or 1254, to insure correct fit for your parts. Generally the type number immediately

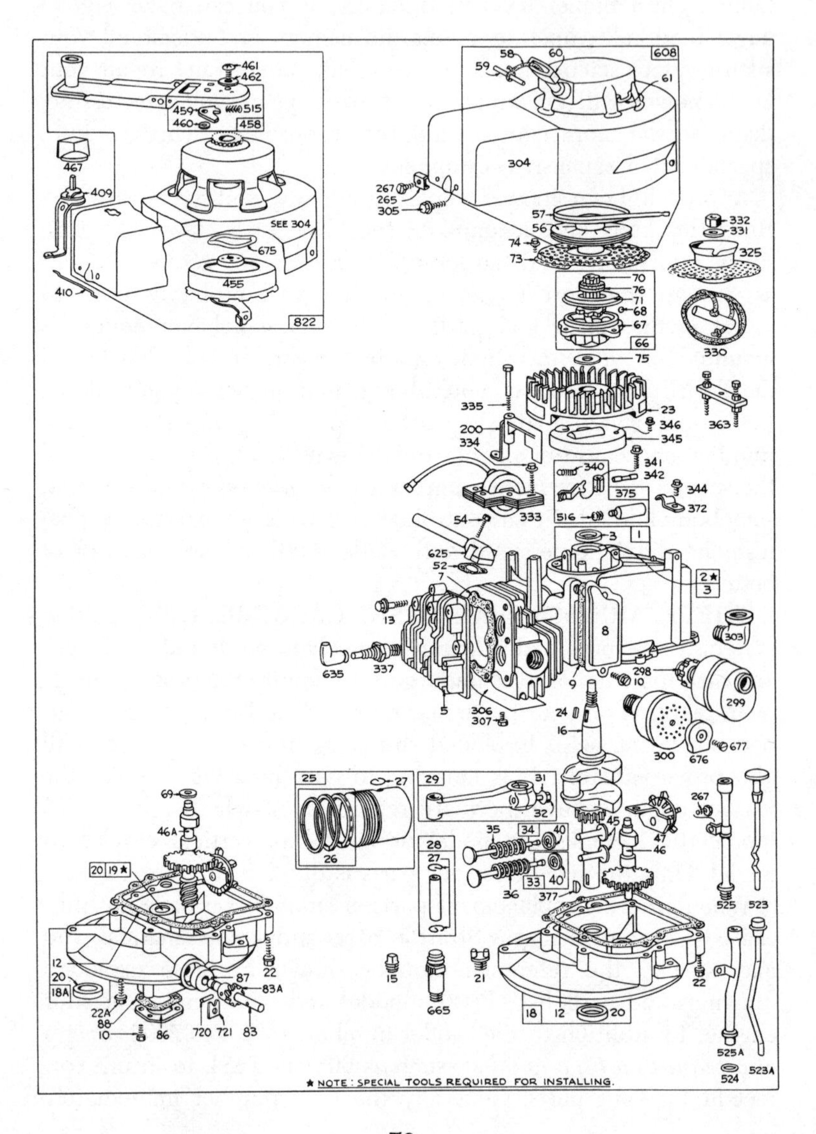

461
462
459
515
460
458
467
409
SEE 304
675
10
410
455
822
608
58
60
59
61
304
267
265
305
57
56
74
73
332
331
325
70
76
71
68
67
66
330
75
335
23
346
345
363
200
334
340
341
342
375
344
372
333
516
54
1
3
625
52
2★
3
7
13
8
303
635
337
298
10
9
299
306
5
307
24
16
300
676
677
25
27
26
29
31
32
267
34
40
35
45
47
46
28
27
33
40
36
377
69
46A
20
19★
12
20
18A
22
87
83A
22A
88
10
86
720
721
83
15
665
21
22
18
12
20
525
523
525A
523A
524
★ NOTE: SPECIAL TOOLS REQUIRED FOR INSTALLING.

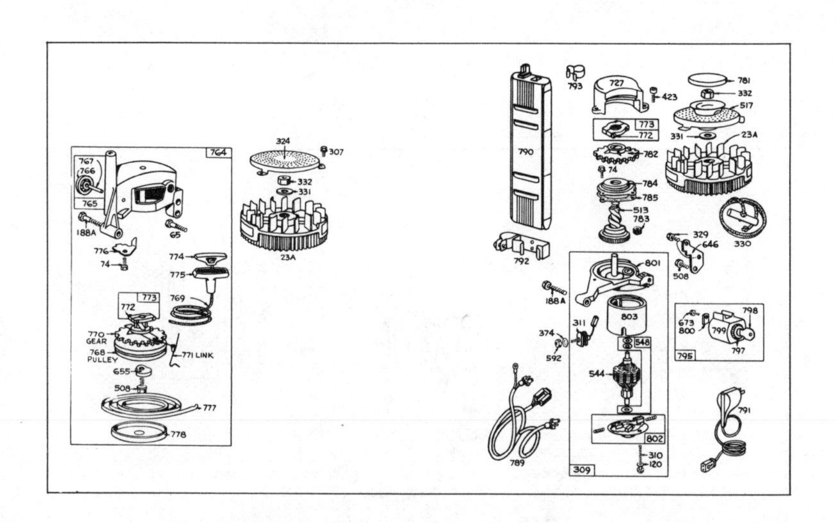

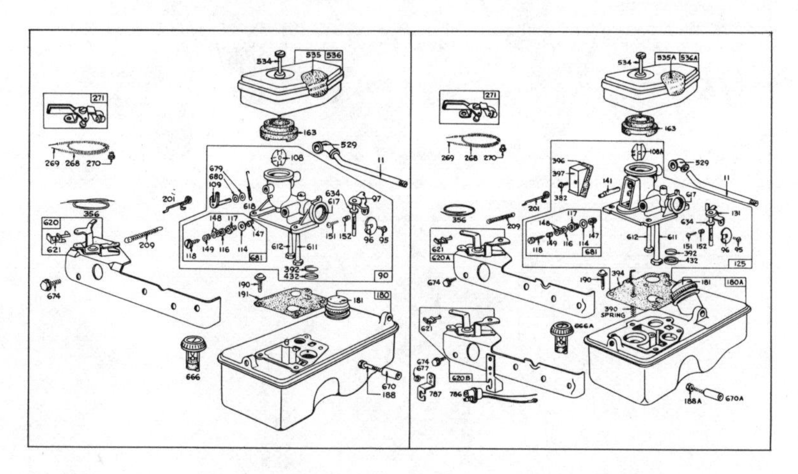

Illustration 4/6. (A-D) A 3.5 h.p. Briggs and Stratton Engine. The insert to Illustration 4/6A shows the optional wind-up starter while Illustration 4/6B pictures an exploded view of the vertical pull starter (left) and battery powered electric starter (right). Types of carburetors which may be included are shown in Illustration 4/6C and Illustration 4/6D.

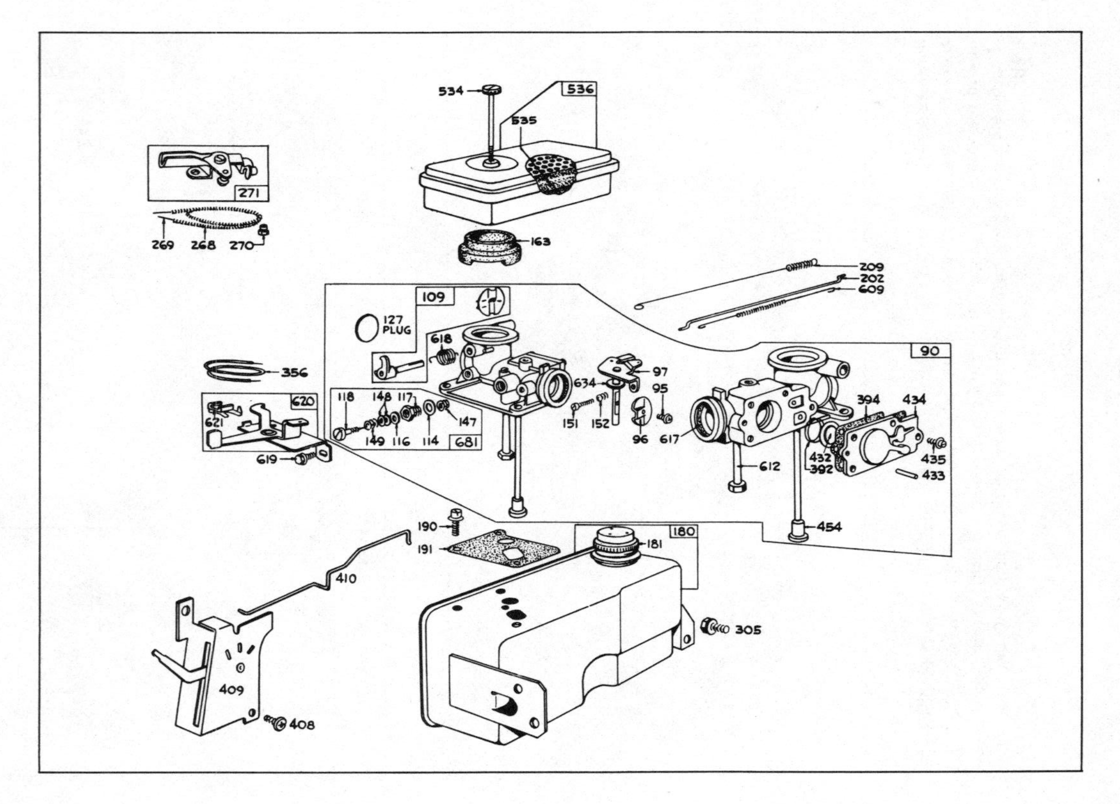

534
536
535
271
269
268
270
163
209
202
609
109
127
PLUG
618
356
620
621
619
118
148
117
147
149
116
114
681
634
97
95
151
152
96
617
90
394
434
432
392
435
433
612
454
190
191
180
181
410
305
409
408

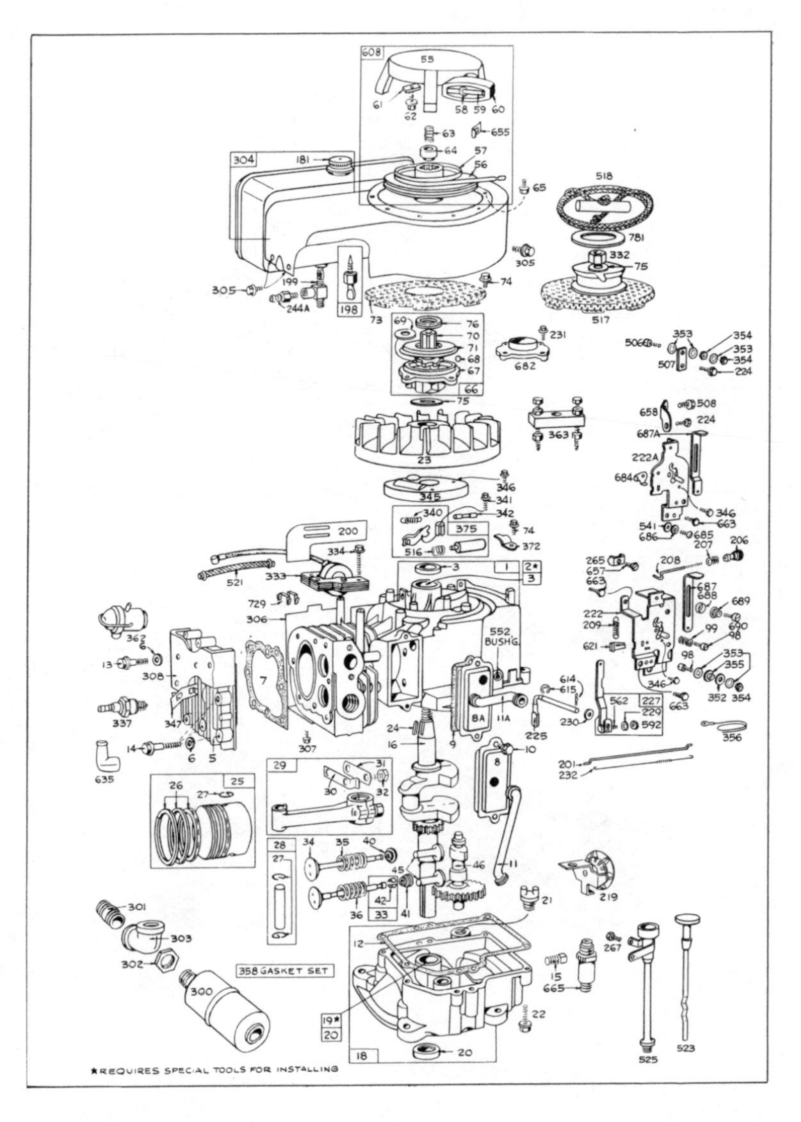

Illustration 4/7A, B & C. A 7 h.p. Briggs & Stratton Engine. Exploded view of this engine resembles very closely the earlier model, the 3.5 h.p. Illustration 4/7B shows the carburetion systems while Illustration 4/7C gives an exploded view of optional electric starters.

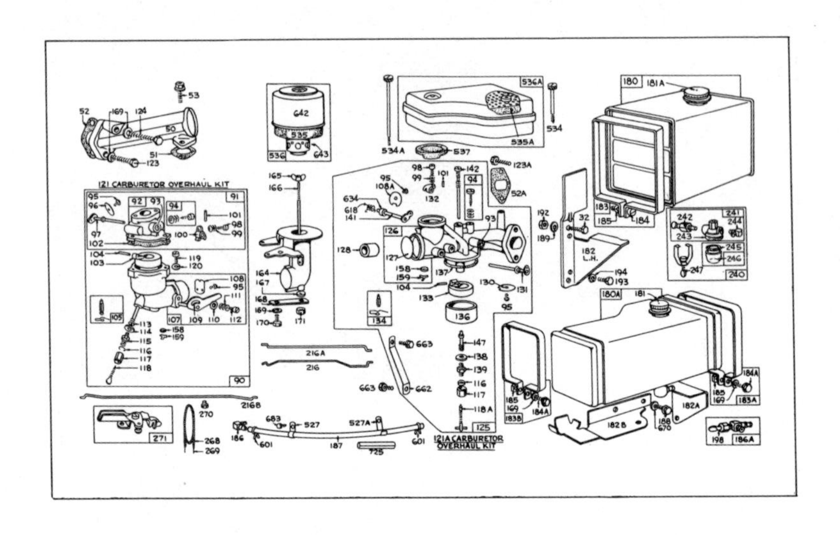
121 CARBURETOR OVERHAUL KIT
121A CARBURETOR OVERHAUL KIT
182 L.H.

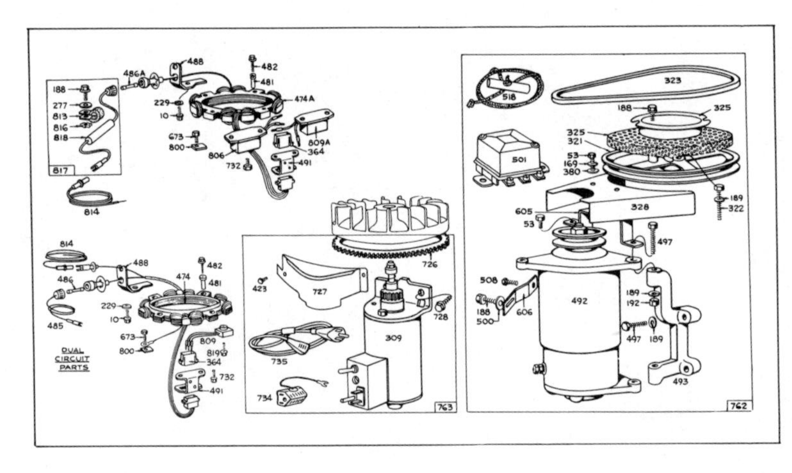
DUAL CIRCUIT PARTS

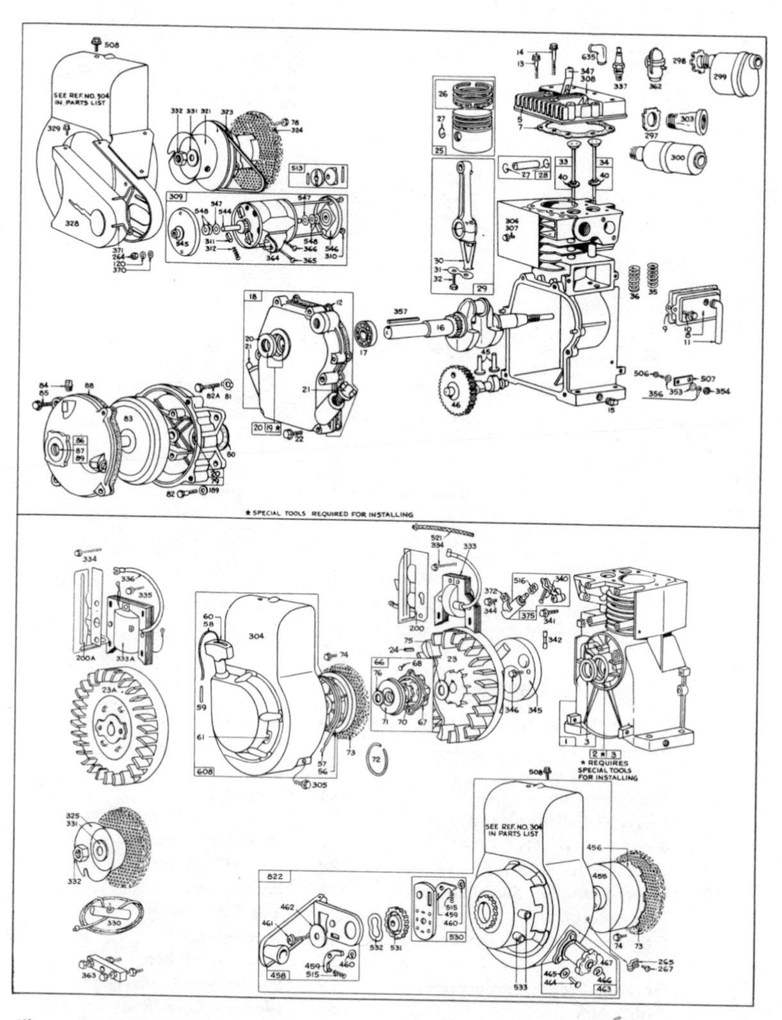

Illustration 4/8A, B & C. A 2 h.p. Horizontal Crankshaft Briggs & Stratton Engine with optional equipment.

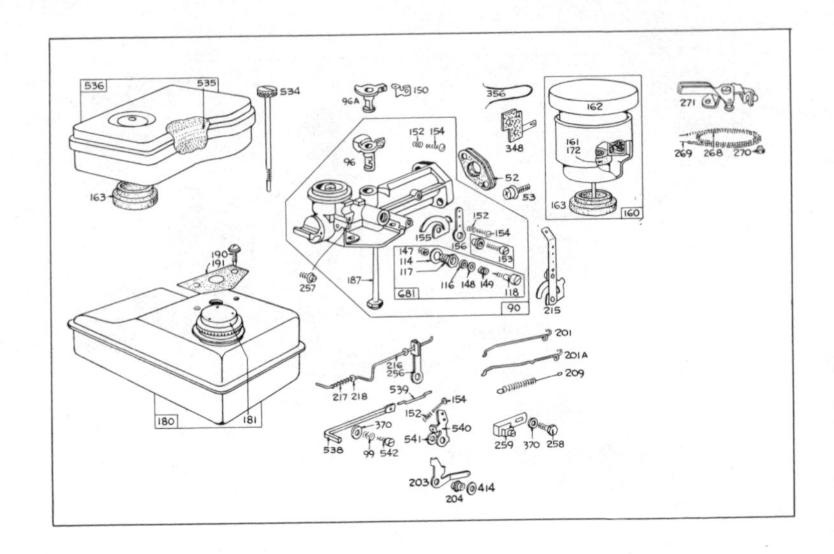

Illustration 4/9 Parts List for Briggs & Stratton Engines

Ref. No.	*Name of Part*
1	Cylinder Assembly
3	Seal—Oil
5	Head—Cylinder
6	Washer—Plain
7	Gasket—Cylinder Head
8	Breather Assembly
9	Gasket—Valve Cover
10	Screw—Breather Mounting Sem
11	Tube—Breather
12	Gasket—Crankcase Cover—1/64″ thick
	Gasket—Crankcase Cover—.005″ thick
	Gasket—Crankcase Cover—.009″ thick
13	Screw—Cylinder Head (2-9/32″ long)
14	Screw—Cylinder Head (2-21/32″ long)
15	Plug—Oil Drain
16	Crankshaft

Ref. No.	*Name of Part*
18	Sump—Oil
19	Bushing—Oil Sump
20	Seal—Oil
21	Plug—Oil Filler
22	Screw—Sump Mtg. Sem
23	Flywheel—Magneto
24	Key—Flywheel
25	Piston Assy.
26	Ring Set—Piston
27	Lock—Piston Pin
28	Pin Assy.—Piston
29	Rod Assy.—Connecting
30	Dipper—Conn. Rod
31	Lock—Conn. Rod Screw
32	Screw—Conn. Rod
33	Valve—Exhaust
34	Valve—Intake
35	Spring—Intake Valve
36	Spring—Exhaust Valve
40	Retainer—Intake Valve
41	Rotocoil—Exhaust Valve

Illustration 4/9 (Continued)

Ref. No.	Name of Part
42	Retainer—Exhaust Valve Rotocoil
44	Pin—Exhaust Valve Rotocoil Spring Retainer
45	Tappet—Valve
46	Gear—Cam
47	Oil Slinger Gear & Bracket
52	Gasket—Intake Elbow Mounting
53	Screw—Carburetor Mounting Sem
54	Screw—Intake Elbow
55	Housing—Rewind Starter
56	Pulley—Rewind Starter
57	Spring—Rewind Starter
58	Rope—Rewind Starter (63" long)
59	Pin—Starter Grip
60	Grip—Starter Rope
61	Bumper—Starter Pulley
62	Screw—Pulley Bumper Mounting
63	Spring—Ratchet
64	Adapter—Ratchet Spring
65	Screw—Rewind Starter Housing Mtg. Sem
66	Clutch Assembly—Rewind Starter
67	Housing—Rewind Starter Clutch
68	Ball—Clutch
69	Washer—Starter Clutch, Thrust
70	Ratchet—Rewind Starter Clutch
71	Washer—Retainer
72	Spring—Clutch Retainer
73	Screen—Rewind Starter
74	Screw—Rotating Screen Mounting Sem
75	Washer—Spring
76	Washer—Ratchet Sealing
78	Screen—Pulley Screen Mounting Sem
79	Case Assembly—Gear
80	Gasket—Gear Case
81	Lock—Counterweight Screw
82	Screw—Cap, Hex. Hd.—5/16-24 x 7/8"
82A	Screw—Cap, Hex. Hd.—5/16-24 x 1-1/2"
83	Shaft Assembly—Drive
84	Plug—Breather
85	Screw—Gear Case Mounting Sem
86	Cover Assembly—Gear Case
87	Seal—Oil
88	Gasket—Gear Case Cover
89	Plug—Oil Drain
90	Carburetor Assembly
93	Bushing—Throttle Shaft
94	Valve Assembly—Carburetor Idle
95	Screw—Throttle and Choke Valve Mtg. Sem
96	Throttle—Carburetor
98	Screw—Machine, Fill. Hd. 8-32 x 5/8"
99	Spring—Throttle Adjustment
101	Pin—Throttle Stop
104	Pin—Float Hinge
108	Valve Choke
109	Shaft and Lever—Choke
110	Washer—Choke Lever
111	Spring—Choke Lever
112	Screw—Choke Lever
114	Gasket—Needle Valve
116	Packing—Needle Valve
117	Nut—Needle Valve Packing
118	Valve—Needle
120	Washer—Lock
121	Carburetor Overhaul Kit
123	Screw—Carburetor Mtg.
125	Carburetor Assembly (Manual Choke)
126	Body Assembly—Carburetor
127	Plug—Welch
128	Venturi—Carburetor
130	Valve—Throttle
131	Shaft and Lever—Throttle
132	Stop—Throttle
133	Float—Carbureator
134	Valve—Fuel Inlet
136	Bowl—Float
137	Gasket—Float Bowl
138	Washer—Float Bowl
139	Retainer—Jet Needle
141	Shaft and Lever—Choke
142	Nozzle—Carburetor
147	Screw—Nozzle
148	Washer—Needle Valve (2)

Illustration 4/9 (*Continued*)

Ref. No.	Name of Part
149	Spring—Needle Valve
150	Connector—Governor Link
151	Screw—Machine, Rd. Hd.—5-40 x 5/8"
152	Spring—Throttle Adjustment
152A	Spring—Speed and Throttle Adjustment
153	Screw—Carburetor Cam and Lever
154	Screw—Machine, Rd. Hd.—5-40 x 5/8"
155	Cam—Speed Adjuster
156	Lever—Speed Adjuster
158	Filter—Carburetor Drain
159	Retainer—Carburetor Drain Filter
160	Cleaner Assembly—Air
161	Bowl—Air Cleaner
162	Cover and Filter—Air Cleaner
163	Gasket—Air Cleaner Mounting
171	Nut—Hex.—1/4-28
172	Gasket—Air Cleaner
180	Tank Assembly—Fuel (1 quart)
181	Cap—Fuel Tank
186	Connector—Fuel Pipe
187	Pipe—Fuel
189	Washer—Lock—5/16 x 1/8 x 1/16"
190	Screw—Fuel Tank Mounting Sem
191	Gasket—Fuel Tank Mtg.
198	Valve—Fuel Shut-Off
199	Outlet—Fuel Tank (Includes Strainer)
200	Guide—Air
201	Link—Governor
203	Crank—Bell
204	Bushing—Governor Lever
206	Nut—Gov. Control Rod
207	Spring—Control Rod
208	Rod—Governor Control
209	Spring—Governor
215	Control—Governor
216	Link—Choke
217	Spring—Choke Link
218	Washer—Choke Link
219	Oil Slinger, Gov. Gear and Bracket Assembly
222	Plate—Gov. Control

Ref. No.	Name of Part
225	Shaft—Gov. Lever
227	Lever Assy.—Governor
229	Washer—Gov. Lever
230	Washer—Gov. Lever
231	Screw—Sem
232	Spring—Governor Link
244	Connector—Fuel Pipe
256	Lever—Speed Adjuster
257	Screw—Fil. Hd., 10-32 x 3/8"
258	Screw—Sem
259	Bracket—Control Wire Casing
261	Screw—Sem
264	Nut—Hex. No. 10-32
265	Clamp—Casing
267	Screw—Clamp Mounting
268	Casing—Control Wire—72" long
269	Wire—Control—78" long
270	Locknut—Control Wire Casing
271	Lever Assy.—Control
297	Locknut—Exhaust Elbow
298	Locknut—Muffler
299	Muffler—Exhaust (Lo-Tone)
300	Muffler—Exhaust
301	Nipple—Exhaust
302	Locknut—Muffler and Elbow
303	Elbow—Muffler—45°
304	Housing—Blower (with Tank)
305	Screw—Blower Housing Mounting Sem
306	Shield—Cylinder
307	Screw—Cylinder Shield Mounting Sem
308	Cover—Cylinder Head
309	Motor—Starting
310	Bolt—Case
311	Brush—Starter Motor
312	Spring—Brush
321	Pulley—Starter
323	Belt—Vee
324	Screen—Electric Starter Pulley
325	Pulley with Screen—Rope Starter (1/2" hole in Pulley)
328	Guard—Belt
329	Screw—Belt Guard Mounting Sem
330	Rope—Starter
331	Washer—Flywheel (1/2" hole)

Illustration 4/9 (*Continued*)

Ref. No.	Name of Part	Ref. No.	Name of Part
332	Nut	460	Retainer—Ratchet Pawl
333	Armature Assembly	461	Screw—Handle Mounting
334	Screw—Armature Mounting Sem	462	Washer—Handle Mounting
335	Screw—Armature Mtg. Sem	463	Holder Assy.—Flywheel
336	Cable—Ignition	464	Rivet—Holder Assy. Mtg.
337	Spark Plug	465	Washer—Holder Assembly Mounting
339	Breaker Assy.—Ignition	466	Retainer—Holder Control Knob
340	Spring—Breaker Arm	467	Knob—Control
341	Screw—Breaker Arm Mounting Sem	506	Screw—Machine, Fil. Hd.—8-32 x 1/2″
342	Plunger—Breaker Point	507	Insulator
343	Condenser	508	Screw—Blower Housing Mounting Sem
344	Screw—Condenser Clamp Mounting Sem	513	Clutch Assembly—Electric Starter
345	Cover—Breaker Point and Condenser	515	Spring—Ratchet Pawl
346	Screw—Dust Cover Mounting	516	Spring—Connector
347	Switch—Stop	517	Pulley—Rope Starter (with screen)
348	Plate Assembly—Stop Switch Insulator	518	Rope—Starter
352	Washer—Insulating	521	Shielding—Ignition Cable
353	Washer—Lock—Shakeproof (3)	523	Cap with Dipstick—Oil Filler
354	Nut—Hex.—8-32 (2)	524	Seal
355	Collar—insulating	525	Tube—Oil Filler
356	Wire—Ground	527	Clamp
357	Key—3/16″ Square	529	Grommet—Breather
358	Gasket Set	530	Housing Assembly—Pawl
362	Shield—Spark Plug (with stop switch)	531	Ratchet—Top Starter
363	Puller—Flywheel	532	Washer—Spring
364	Plug—Connector	533	Knob—Stop
365	Screw—Plug Mtg. (long)	534	Screw—Air Cleaner
366	Screw—Plug Mtg. (short)	535	Element—Air Cleaner
370	Washer—Plain	536	Cleaner Assembly—Air
371	Restrictor—Belt	537	Gasket—Air Cleaner
372	Clamp—Condenser	538	Rod—Speed Control (square rod)
375	Breaker Points and Condenser Set	539	Link—Control Rod (Used only with square rod)
392	Spring—Fuel Pump Diaphragm	540	Lever—Speed Adjuster
394	Diaphragm	541	Washer—Speed Adjuster to Carburetor
414	Washer—Bell Crank		
423	Screw—Sem	542	Screw—Machine, Fil. Hd.—10-32 x 3/4″
432	Cap—Spring		
455	Cup, Hub and Spring Assy.	544	Armature—Starter
456	Plate—Retainer	545	Case Assembly—Long Shaft End
458	Handle Assy.—Cranking	546	Case Assembly—Short Shaft End
459	Pawl—Ratchet	547	Washer—Fiber

Illustration 4/9 (Continued)

Ref. No.	Name of Part	Ref. No.	Name of Part
548	Washer—Thrust (Steel)	679	Washer—Choke Shaft (Felt)
552	Bushing—Governor Shaft (no threads)	680	Washer—Choke Shaft
		681	Needle Valve Kit
552A	Bushing—Governor Shaft (with threaded end)	682	Shield—Starter Clutch
		683	Screw
562	Bolt—Governor Lever	684	Stop—Control Lever
590	Screw—Breaker Mounting Sem	685	Screw—Machine, Rd. Hd.—5-40 x 1/2″
592	Nut—Hex., 10-24		
601	Clamp—Fuel Pipe	686	Washer—Lock
608	Starter Assy.—Rewind	687	Slide—Friction Control
609	Spring—Throttle Link Used on type No. Series 9000 to 9200 and 952010 to 952999.	688	Cap—Friction Spring
		689	Spring—Friction
		690	Screw—Machine, Fil. Hd.—10-32 x 1/2″
611	Pipe—Fuel, (Long) Plastic		
612	Pipe—Fuel, (Short) Plastic	696	Key—Timing Gear
614	Cotter—Hair Pin	720	Pin—Gear to Shaft
615	Retainer—E-Ring	721	Stop—Drive Shaft
617	Seal—Intake Tube	726	Gear, Ring (Includes mounting parts)
618	Spring—Choke Return		
620	Plate—Carburetor Control	727	Cover and Shield—Starter
621	Switch—Stop	728	Screw—Sem
625	Tube—Fuel Intake	729	Clip—Wire
634	Washer—Throttle Shaft (Felt)	734	Plug—Adapter
635	Elbow—Spark Plug	735	Cord—Starter (includes Adapter Plug)
646	Bracket—Electric Starter Mounting		
655	Anchor—Spring	741	Gear—Timing
657	Screw—Sem	757	Link—Counterweight
658	Strap—Control Plate	758	Counterweight Assembly
659	Nut—Governor Shaft Bushing	759	Pin—Counterweight (2)
660	Bracket—Control Plate	760	Spacer—Counterweight (2)
662	Bracket—Carburetor	761	Screw—Counterweight (2)
663	Screw	763	Electric Starter Kit—110 Volt (Gear Drive)
665	Oil Minder		
666	Gauge—Fuel Level Used only on type Nos. 0110 to 0300.	764	Starter Assembly—Vertical Pull
		765	Housing—Vertical Pull Starter
		766	Pulley—Rope (Small)
666A	Gauge—Fuel Level Used only on type Nos. 0600 and Up.	767	Pin—Small Rope Pulley
		768	Pulley—Rope Gear
		769	Rope—Vertical Pull Starter
670	Spacer—Fuel Tank	770	Gear—Vertical Pull Starter
670A	Spacer—Fuel Tank	771	Clip—Vertical Pull Starter
673	Screw—Receptacle Retainer	772	Screw—Sem
674	Screw—Tank to Bracket	773	Retainer Assembly—Spur Gear
675	Washer—Starter Housing Spring	774	Insert—Starter Handle
676	Deflector—Muffler	775	Grip—Starter Rope
677	Screw—Muffler Deflector and Harness Retainer	776	Guide—Rope
		822	Starter Assy.—Wind-up

follows the model number in all labels and lawn mower parts lists. Absolutely precise nomenclature, therefore, would be "Model 92902 Type 0061."

In the drawings which show exploded views of Briggs & Stratton engines, every part in each illustration is labeled with a *reference number* (ref. no.). Parts which are identical in various engines, aside from a difference in overall size, have identical reference numbers. In the relatively small engine Model 92599 and in the relatively large engine Model 171706, the two flywheels are of considerably different sizes. But they look identical in the drawings and serve the identical function. They therefore have identical reference numbers, in the case of these flywheels, (ref. no. 23).

Piston rings, for example, come in a variety of sizes but they all have (ref. no. 26). Mufflers are generally (ref. no. 299) or (ref. no. 300). It is obvious, therefore, that reference numbers are useful for studying exploded diagrams but almost useless for ordering parts of the right size. A muffler (ref. no. 300) for a Model 171706 engine bears part number 294599 whereas (ref. no. 300) for a Model 19599 engine has been given the part number of 298830.

BRIGGS & STRATTON IGNITION. One of the most persistent complaints which seasoned repairmen make about the smaller Briggs & Stratton engines is that their ignition parts are mounted behind the flywheel. To get at the magneto armature, points, condenser and breaker-plunger, a repairman first must remove the flywheel. A veteran repairman with a well-stocked machine shop can loosen a flywheel and start servicing the ignition in 10 minutes. A novice should proceed as slowly as necessary, however, and refrain from the temptation of coaxing the flywheel loose with a hammer. It may be rugged, but it *can* break or warp.

Aside from routine maintenance, your first engine repair job may very well be in the ignition system, which means you'll be pulling off a flywheel early in your repair apprenticeship. Patience and paying attention to step by step instructions will get both you and your lawn mower engine through safely. Unless you happen to be near a well stocked Briggs service station, before digging very deeply into ignition you'll want to buy a flywheel key (ref. no. 24) and an ignition kit (ref. no. 375) before starting ignition repairs.

If your mower has a simple rope starter (engine model code 1), or an electric starter (engine model code 4, 5 or 7), a 1/2-inch wrench is probably best for removing the one nut which secures the flywheel; a socket wrench is preferred. The nut generally will have a left-hand thread; in other words, you will have to twist it to the right to loosen it, the opposite direction from most parts which are screwed together.

Rewind and windup starters (engine model code 2, 6 or 8), are secured in place by the starter clutch housing (ref. no. 67). Briggs makes a special starter clutch wrench (part number 19114) which slips easily over the four notches in the starter clutch and enables you to unscrew it with little difficulty. Lacking such a special wrench, you'll have to improvise. The opened jaws of a pliers generally suffice although, with luck, the handle end of your pliers might be close to the very size tool required.

By whatever means you twist off the retaining nut or clutch, place a block of wood against the vanes of the flywheel so it cannot turn while you remove the nut or the starter clutch housing. In figure 4/11 you'll see yet another Briggs refinement, part number 19167, *a flywheel holder.* This does away with the need for a block of wood beneath the flywheel while you twist off the retaining nut or starter clutch housing.

Once the nut or clutch housing is off, the flywheel itself is still held onto the crankshaft by friction and maybe even some rust or gum. To make life easy for repairmen and to safeguard against damaging or warping the flywheel by somebody's over-exertion, Briggs and other manufacturers make an inexpensive flywheel puller which is shown in illustration number 4/11. Local hardware stores might stock a large gear puller costing more than the Briggs specialized flywheel puller, which not only could slip the flywheel off effortlessly but also would be valuable for other chores. The less expensive Briggs & Stratton flywheel puller, although quite modest in price, is of value only on Briggs & Stratton flywheels—and then just for specific flywheels. Without any of these tools, pry gently beneath the flywheel, being careful not to pry near the magnets under there. Once the flywheel is off, your troubles are practically over.

Many repair manuals issued by engine manufacturers other than Briggs & Stratton recommend another way to loosen the flywheel. A mechanic is instructed first to remove the bolt or

Illustration 4/10. Removing (or Re-Installing) a Flywheel. A block of wood steadies the flywheel, making removal of the nut or clutch which holds it in place easier.

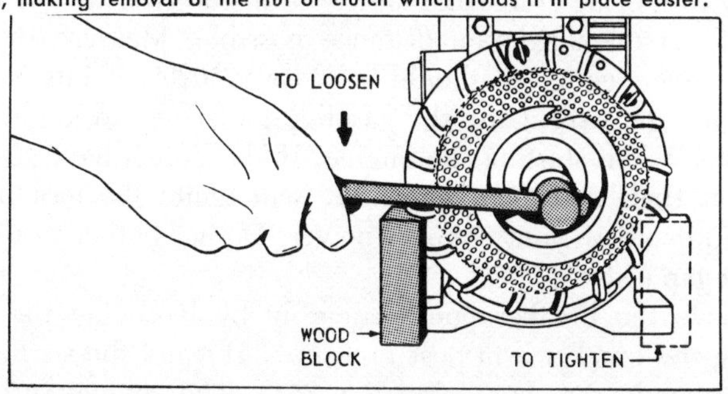

Illustration 4/11. Using a Flywheel Puller. Briggs & Stratton, like most engine manufacturers, offers for sale a tool which makes removing the flywheel from the shaft an easier and safer job.

clutch housing which secures the flywheel to the crankshaft. Then he screws on a "shock nut" which both protects the crankshaft threads and serves as an anvil upon which the mechanic can pound with relative safety. Although the shock nut is very inexpensive, if one is unavailable when you need it, a substitute can be fashioned by protecting the crankshaft with two or three standard nuts. With the shock nut in place, a mechanic gives the nut (and the crankshaft at the same time) a sharp rap with a hammer. The resulting shock jars the crankshaft and flywheel loose from each other.

With the flywheel set aside, the breaker cover (ref. no. 345) is now exposed. The screws holding down the cover are removed and the cover gently pried loose. If the cover is damaged in any way or bent, order a new one. The purpose of the breaker cover is to protect the ignition breakers underneath from dust and other foreign ingredients which would foul your ignition system. Before putting the cover aside, inspect the inside for signs of oil. After a year or two without being removed, an extremely thin film of oil spread evenly over the cover might be expected, but a heavy, uneven film, or even drops of oil, can be caused by a worn breaker point plunger (ref. no. 342).

Before removing the breaker points, rotate the crankshaft several times while watching the points. They should move solidly together and then apart while you rotate the crankshaft. Finally rotate the crankshaft slowly until the two breaker points separate to the maximum distance possible. Measure that gap with a feelers gauge. It should be close to .020″. Points with too much of a gap will ignite the gasoline too soon and might cause kick-back when starting the engine. If the points have too little space at their maximum gap, they will ignite the gasoline too late and decrease your engine's power. Do not bother to readjust the old gap just yet.

Remove the breaker point assembly by loosening the screw holding the breaker arm post (ref. no. 341) and the screw holding the condenser clamp (ref. no. 344). The contact surface of each point should be smooth, unburnt and unpitted and each point should be at least 1/32″ thick. It hardly pays to reinstall the old set of points since a kit of new ignition parts is so inexpensive. Two such kits are available. One kit contains points, condenser and related hardware. The second kit, in addition to points and condenser, contains a new breaker point plunger (ref. no. 342) and a new flywheel key (ref. no. 24).

Once the breaker points and condenser are removed, the plunger end is exposed. Examine it closely for evidence of oil leaks. Oil on the breaker points themselves causes misfiring and carbon deposits. The two sources for oil inside the breaker cover area will be through the breaker plunger hole or through the oil seal around the crankshaft. If in doubt, replace both if oil fouling is a problem. The oil seal replacement, covered further on, is easier to accomplish than installing a new bushing in the oil plunger hole. These bushings are generally found only in larger engines.

The crankshaft must be removed before the breaker plunger bushing can be replaced. That procedure is described later. To accomplish the job, three modestly priced special tools are on sale at Briggs central service distributors: two reamers (part number 19056 and 19058) plus a bushing driver (part number 19057).

The smaller of the two special reamers, number 19056, is used to remove all traces of the old bushing. Metal scraps and shavings must be meticulously cleaned away to prevent their damaging the crankshaft or bearings or other moving parts. Then the new bushing is placed over the hole and driven down with the 19057 bushing driver. The top of the new bushing must end up flush with the top of the breaker plunger hole. Briggs' special tool is useful simply to avoid damaging the bushing as you install it. Briggs recommends that you finish off the job with their 19058 reamer, a task you could very well handle by judicious application of a small file or emory cloth.

Ironically, the three special bushing tools cost many times more than the very inexpensive bushing itself. As an improvised alternative, a handyman could use taps of the appropriate size in lieu of the reamers. A dowel stick or two can drive the bushing into place.

NOW LET'S PUT THE IGNITION SYSTEM BACK TOGETHER. Any parts which are going to be reinstalled instead of replaced by new ones must be cleaned thoroughly. A good grade of kerosene cleans nicely but there are several proprietary products intended specifically for cleaning engine parts which will work a shade better.

Insert the breaker point plunger into its appropriate hole. A bit

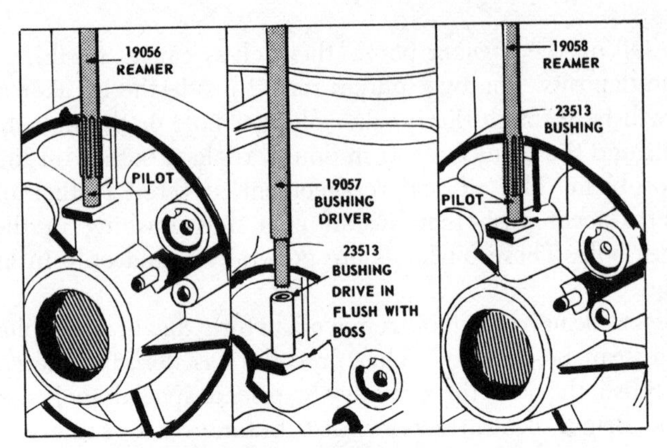

Illustration 4/12. Installing Breaker Plunger Bushing.

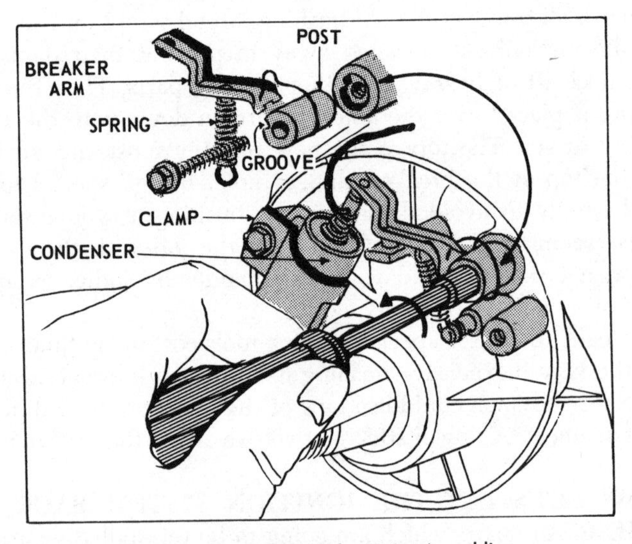

Illustration 4/13. Breaker Point Assemblies.

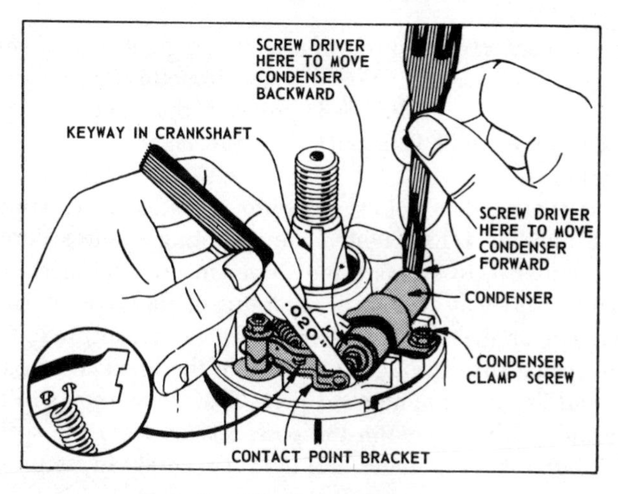

Illustration 4/14. Adjusting Breaker Point Gap.

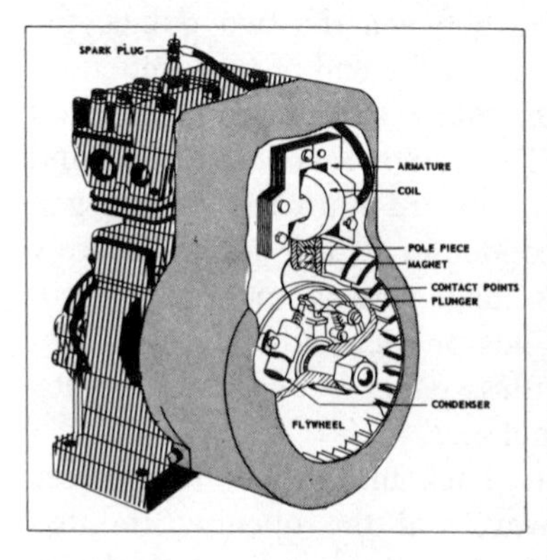

Illustration 4/15A.
Flywheel Ignition Internal Breaker.

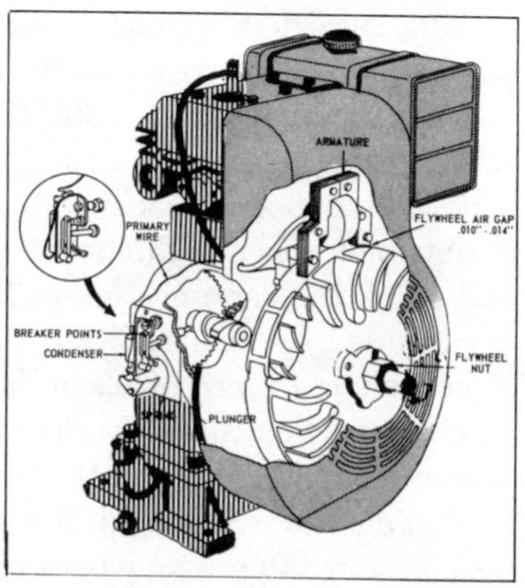

Illustration 4/15B.
Flywheel Ignition External Breaker.

of oil may be necessary to slip it in easily. If a great deal of pressure is needed to insert the oiled plunger, something probably is amiss. If one end of the plunger is grooved, that end must face the breakers or oil will leak out when the engine is put back into operation.

Next put the breaker points post into place as demonstrated in drawing number 4/13. Tighten the mounting screw securely. Drawing number 4/13 shows a nut driver in use, the most convenient and surest way to install apparatus of this type. A small socket wrench or even a box wrench can be substituted. One end of the spring slips through *two* holes in the breaker arm and the other end fits over the notched post. That done, then slip the breaker arm into the groove in the post; tension from the spring will pull it against the plunger and hold it securely in its proper place.

Attach the primary ignition wire, the one leading to the magneto coil, to the condenser by depressing the spring fastener, inserting the wire, and then releasing the spring again. Put the condenser into its clamp and only loosely screw it into place, making sure that contact is made between the two points, the one on the breaker arm and the one on the end of the condenser.

Turn the crankshaft until the points have opened to their widest possible gap. Insert a feelers gauge opened to .020″ and slide the condenser backward or forward until the proper gap is obtained. Then tighten the condenser clamp screw securely and check the gap once more. To make sure the points have not been fouled by oil from your hands or some other foreign substance, slide a piece of clean, lintless paper between the points, making sure that the paper does not tear.

Screw the breaker point cover back into place, but before tightening the screws all the way, seal the opening through which the ignition wire passes with Permatex or silicone sealer as additional protection against dirt. Finally, turn the two screws down firmly and evenly so as not to distort or damage the cover. If you have any doubts about the cover's soundness, replace it. The cost is very, very little.

Now you can put the flywheel back on after making sure

that it and all related pieces are clean and dry. The key slips into the crankshaft keyway and the spring washer goes on next, hollow side of the washer facing the flywheel. If for any reason you should ever fabricate a temporary key to replace a lost original one, do not use hard steel. The genuine ones are zinc, aluminum or soft iron.

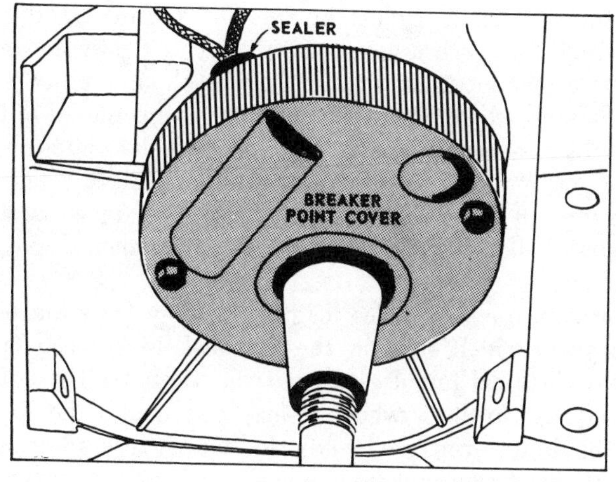

Illustration 4/16.

To secure the flywheel nut or starter clutch housing on the crankshaft, simply reverse previous instructions on how to remove it. Most engine repair manuals specify exactly how tight various bolts ought to be. Briggs & Stratton is no exception. To accomplish such a feat, a *torque wrench* is used. The torque wrench typically attaches to a socket wrench and includes a dial which indicates how much pressure you are exerting on a bolt, pressure being measured in "foot pounds" or "inch pounds." The latter is 1/12th the size of the former. A torque wrench is a valuable addition to toolboxes of repairmen and serious handymen.

Here are the various torque values listed by Briggs & Stratton for tightening flywheel nuts:

Models in the 60000 to 92000 series	57 ft. lbs.
Models in the 100000 and 130000 series	60 ft. lbs.
Models in the 140000 to 190000 series	67 ft. lbs.

It is entirely feasible to reinstall lawn mower engine bolts without a torque wrench. Unless instructions say to do otherwise, tighten each bolt with a good wrench just as tightly as you can *easily* get it. Do not strain against the wrench and do not jerk on it. In situations where more than one bolt has been used to fasten an assembly, concentrate on tightening the bolts uniformly.

When you removed the flywheel in the first place, you probably also took off the governor air vane and the magneto armature coil. Reinstall both of them at this point with their appropriate screws.

The armature coil receives magnetic energy from the permanent magnets which spin on the flywheel. Magnetism is converted to electrical impulses in the coil. If the coil is adjusted to sit too close to the flywheel, it may rub against the spinning metal. Should the armature be adjusted too far from the magnets, the electrical charge will not be strong enough to operate the engine efficiently. Therefore, the gap between the end of your armature's iron core and the edge of your flywheel must be set with significant precision. Fortunately, arriving at the correct magneto gap is not nearly as complicated as it might sound.

Screw holes in the armature are slotted. Shove the armature as far away from the flywheel as it will go and tighten one screw. Select a feelers gauge of the appropriate size and slip it the long way between the armature and flywheel. Now twist the flywheel until the magnets are directly beneath the armature. Loosen the one screw and the magnets should pull the armature down until it rests firmly upon the feelers gauge. Tighten both mounting screws firmly, move the magnets away from the armature, pull out the feelers gauge, and spin the flywheel. It should revolve without scraping any part of the armature.

For model 60000 through 92000 series engines, use a .006″

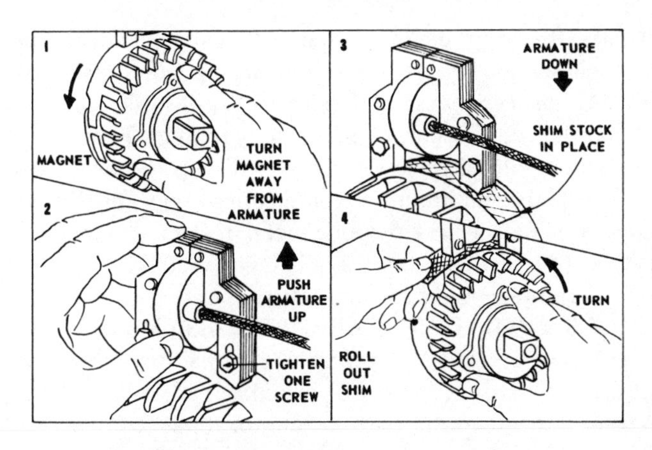

Illustration 4/17. Adjusting Armature Air Gap.

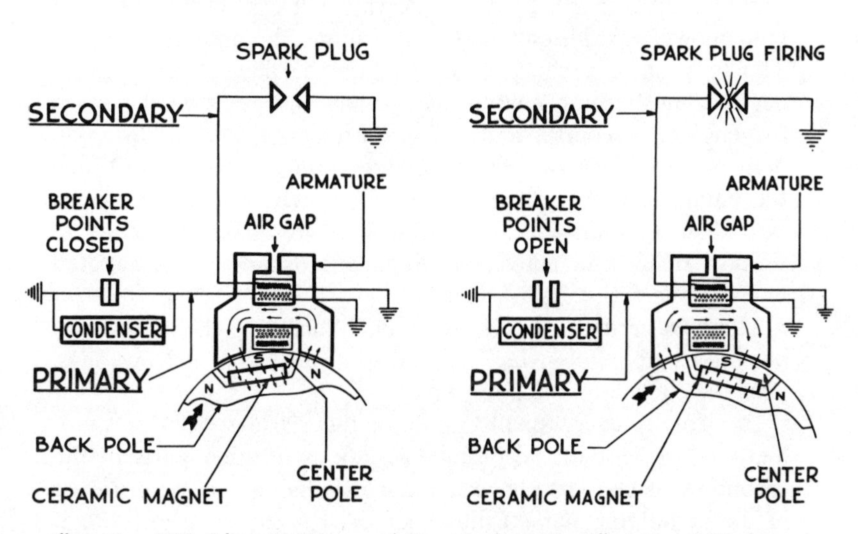

Illustration 4/18. Schematic Diagram of Magneto Operation. Illustration 4/18 shows how the electric charge created by the permanent flywheel magnets passing the armature's coil is grounded because the breaker points are normally closed. When it is time for the spark plug to fire, causing ignition, the breaker points open and the electric charge sparks the gasoline-air mixture, as shown in Illustration 4/19. The condenser merely helps cushion the quick surge of electricity and prevent the points from burning out prematurely.

gauge and for larger engines use a .010″ gauge. The specific gaps recommended by Briggs & Stratton are actually .006–.010″ and .010–.014″ respectively. But Briggs uses a flexible shim (available on order) for accurate magneto gapping. The wings of a feelers gauge, although less flexible, will gap the magneto properly. Select the minimum recommended size for the gap to compensate for the less flexible metal feelers gauge. As an alternative, use "Scotch" tape. Each layer is .0022″. Four or five layers, therefore, could provide an appropriate gap.

BRIGGS & STRATTON CARBURETORS (Model 92000 series engines): Before extensive repair or internal inspection, have on hand a set of gaskets for your particular carburetor. If you suspect wear, buy an appropriate carburetor repair kit.

Remove the air cleaner as described in chapter two under routine maintenance. In doing so, you will expose the throat of your mower's carburetor and can examine the automatic choke. The purpose of a choke is to regulate the flow of air into the engine. The burning mixture of gasoline and air is what powers the engine, of course, and the carburetor must be adjusted to deliver a satisfactory mixture under varying conditions. When the engine is just starting or running slowly, the mixture must be rich in gasoline; a condition which tyros have shortened to simply "rich." And conversely, high speeds are best generated in an engine with a "lean" mixture.

There was a time when every engine, including the ones which powered automobiles, had a hand-controlled choke. Nowadays, even the choke is automatic.

In virtually every small engine of the 92000 series, the automatic choke is identical despite variations in other parts of the carburetor itself. The choke is a flat disc filling the entire *throat* of the carburetor, named thus because it is the passage through which air is drawn. The disc is attached via a thin rod to a diaphragm atop a chamber linked to the cylinder's intake port. As the intake valve opens and the piston begins its intake stroke, the resulting vacuum pulls on the diaphragm which in turn pulls the choke partially open. As the engine speeds up, the piston moves faster and the vacuum exerted on the diaphragm becomes greater, opening the choke still farther. Because of that chain of events, more air is mixed with the gasoline which results in a leaner mixture of fuel going to the cylinder.

The Briggs & Stratton 92900 and 92500 series engines are among the most frequently encountered powerhouses atop rotary lawn mowers. Briggs appears to be phasing out the 92500 with its "Vacu-Jet" carburetor in favor of the 92900 with its "Pulsa-Jet" carburetor. The difference, as far as most repairs are concerned, is quite esoteric. The 92000 carburetor and 100900 carburetors are mounted directly on top of the fuel tank and the machined metal top of the tank is made an integral part of the carburetor.

Remove the tank-carburetor assembly from the engine and then remove the carburetor from the tank. A series of only a few screws holds each in place but proceed slowly and disconnect all cables and tubes from the carburetor and tank as you go along. In removing the carburetor from the gasoline tank, be warned that a flexible diaphragm (ref. no. 394) lies beneath. Pry, if you must, but only at the very edges so as not to tear or disfigure the diaphragm. Inspect the diaghragm for signs of wear or for punctures. Hold it up to the light to locate even tiny holes. If the diaphragm is in perfect condition, it can be used again.

You should check the length of the choke spring (ref. no. 390) beneath the diaphragm. The Pulsa-Jet spring must be a minimum of 1-1/8″ long and a maximum of 1-7/32″. On Vacu-Jet models, the minimum length is 15/16″ and the maximum is one inch. If the spring is too long or too short, replace *both the spring and the diaphragm,* a relatively inexpensive investment.

Since the tank top performs as an integral part of the carburetor interior, the machined portions of it must be very flat. If not, gasoline can leak beyond the diaphragm and into the vacuum chamber. From there it can progress to the carburetor section, causing flooding. Or drops of raw gasoline can enter the cylinders where they wash away the protective oil coating and thus shorten the life of your engine's moving parts.

As shown by illustration number 4/21, lay a good straight edge, such as a new steel ruler, across the top of the machined

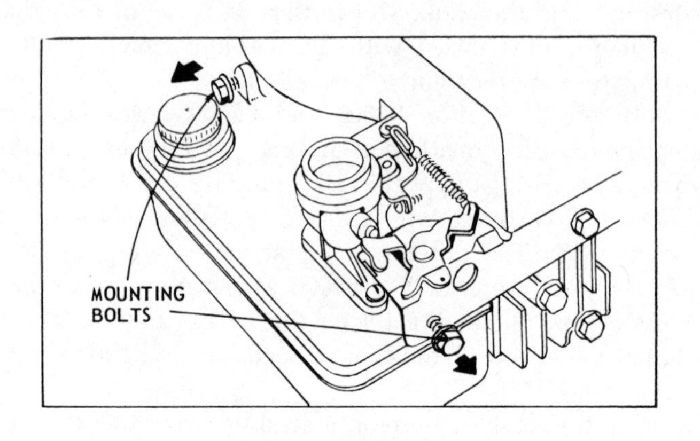

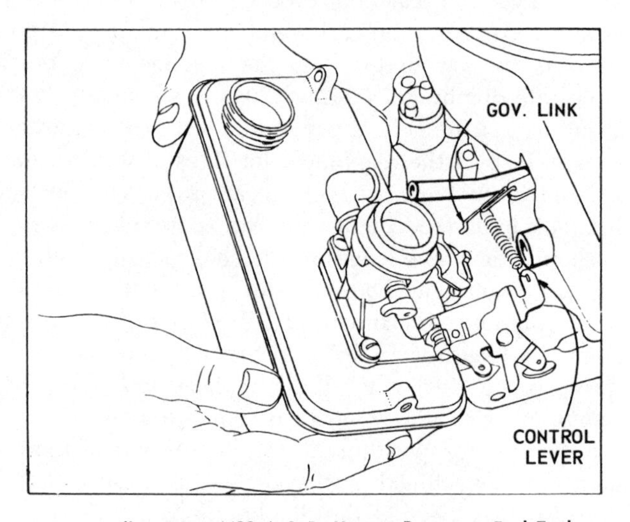

Illustration 4/20 A & B. How to Remove a Fuel Tank.

portion of your fuel tank. Gently try to insert a .002″ feelers gauge between the straight edge and the tank at the areas shaded in the round insert. If there is enough clearance for the .002″ gauge to slip between the ruler and the tank, the tank must be replaced. A shiny new Briggs & Stratton tank (of the proper part number) is not expensive.

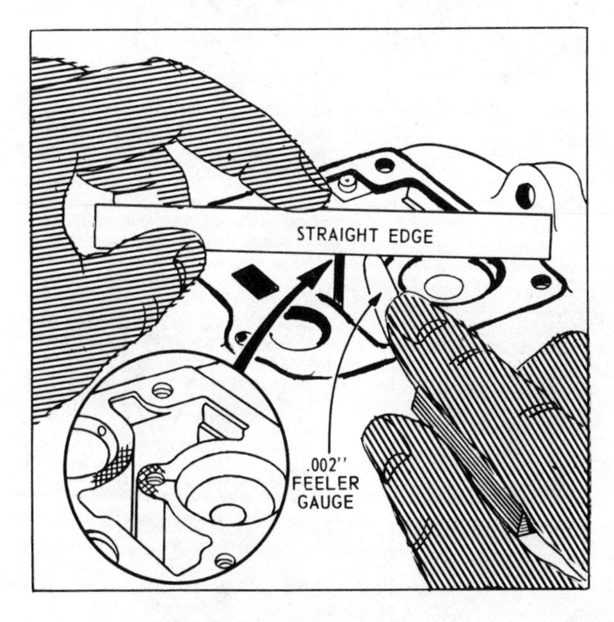

Illustration 4/21. Checking Tank Top Flatness.

If additional repairs are needed, they will be dealt with in a page or so. For the moment, let's spend some time seeing how the diaphragm and choke springs go back together. Figure 4/22 demonstrates the art pictorially. Insert the uncoiled end of your new choke spring through the underside of the new diaphragm. Please don't bend the spring in the process. As you lay the new diaphragm on top of the well cleaned tank surface, make sure that the spring goes into the spring pocket as shown in drawing number 4/22b. Pulsa-Jet carburetors will also have a cap and spring for the fuel pump section of the carburetor which also must be carefully positioned.

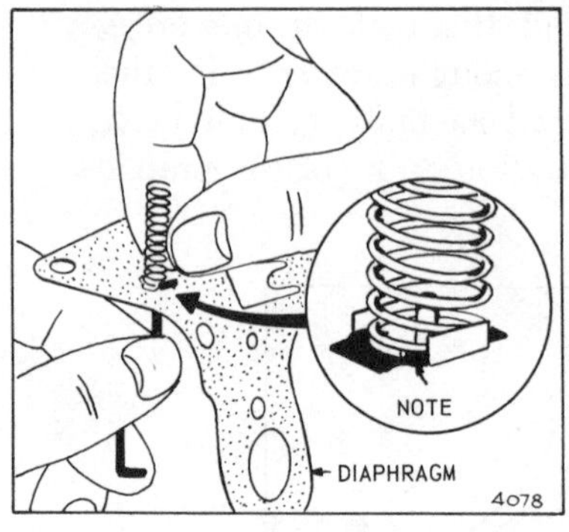

Assemble Spring to Diaphragm

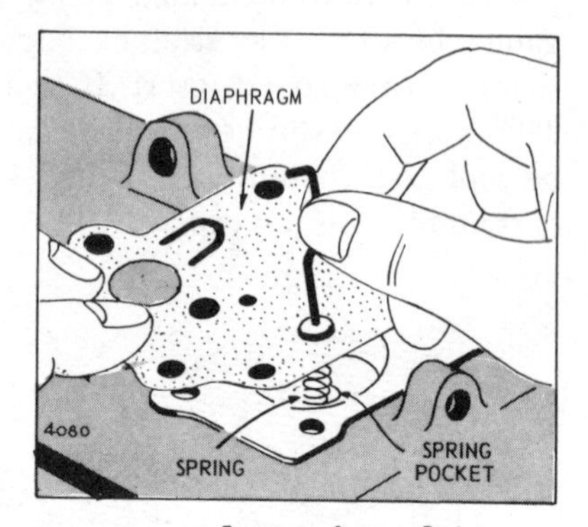

Positioning Spring in Pocket

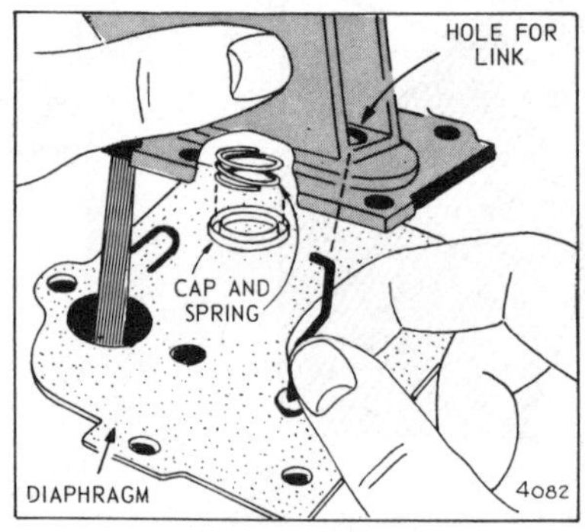

Locating Diaphragm on Tank Top

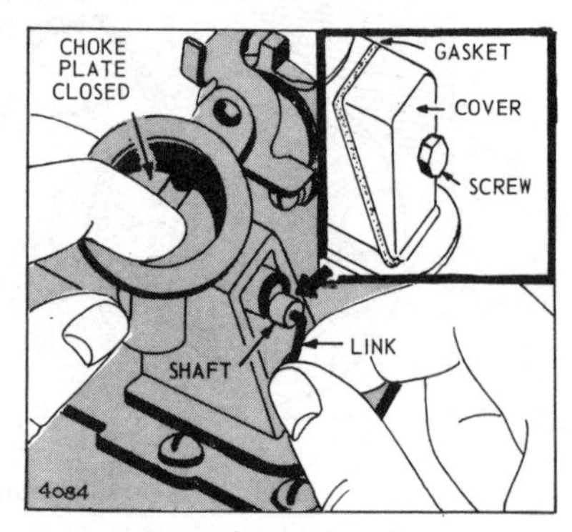

Inserting Choke Link

Illustration 4/22. How to Install a New Diaphragm on Small Carburetors.

Lay the cleaned carburetor onto the diaphragm carefully and thread the mounting screws only about two turns. Close the choke plate with your finger and attach the choke link. Now put the cover back in place, as shown in the insert of drawing number 4/22d. It wouldn't hurt to spend a few pennies for a new gasket (ref. no. 396).

Again with your finger, hold the choke plate *wide open* this time, which is approximately perpendicular from its normally closed position. While the choke plate is being held open, firmly tighten down the carburetor mounting screws. To insure that you do not misshape the diaphragm, first turn the mounting screws in by hand; finally tighten each screw only one-half turn at a time, alternating from one side of the carburetor to the other, until all of them are firmly tightened into place.

Remove your finger from the choke plate and it should move easily to its fully closed position. If not, try it again. Make sure that the choke spring is properly assembled in the diaphragm and is correctly inserted in its pocket within the tank top.

To remove the choke disc itself and related choke parts on the 92500 and 92900 series carburetors, first disconnect the choke return spring at the carburetor end. The choke shaft (ref. no. 141) is nylon and will pull out with relative ease unless it has been heat-sealed to the choke valve (ref. no. 108a). In that event, a pointed tool such as a knife should be run along the edge of the shaft. Clean or replace each of these inexpensive parts.

Since there are few moving parts within carburetors installed on small Briggs & Stratton engines, almost all other repair work on the fuel system consists of cleaning. Fuel pipes which extend from the carburetor body into the gasoline tank are made of nylon. There is a screen built into the bottom of each pipe and there is often a check valve near the bottom which consists of a tiny ball to prevent fuel or air from being forced backwards into the tank. These fuel pipes are removed and installed with a socket wrench or box wrench, generally 9/16″. When putting in a new pipe or replacing the old one, use the socket in your hands rather than in the wrench to avoid over-tightening the pipe

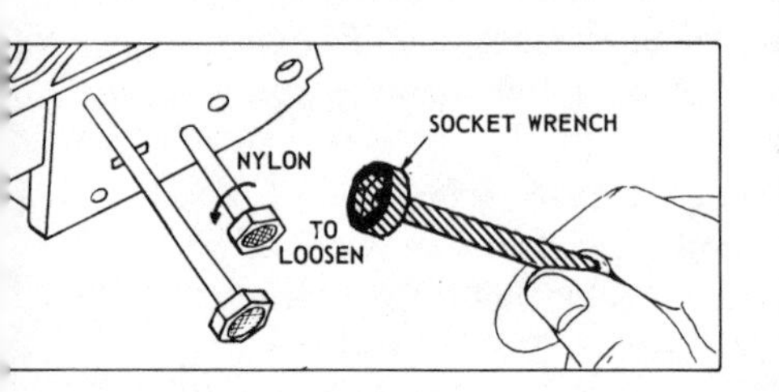

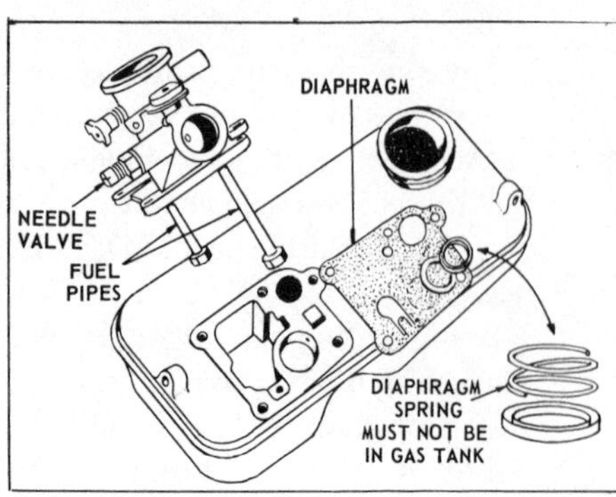

llustration 4/23. Replacing Nylon Fuel Pipes.

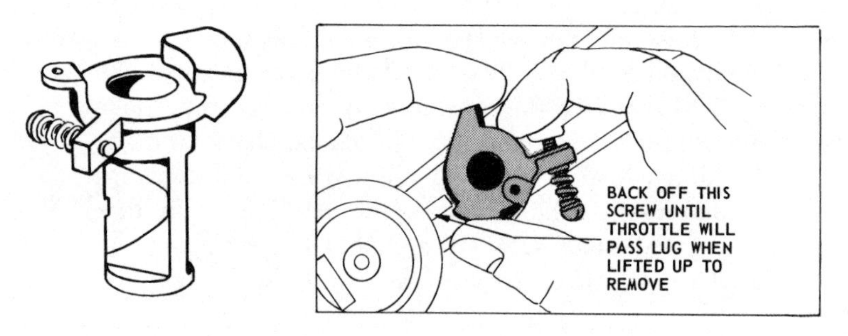

Illustration 4/24. How to Remove Cast Iron Throttles from Carburetor. This type of throttle is found on older engines.

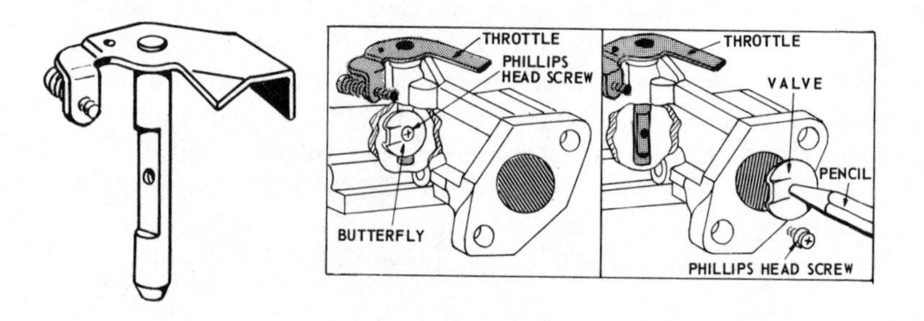

Illustration 4/25. How to Remove Stamped Throttles from Carburetors.

and damaging the threads which are also made of nylon. The long pipe supplies fuel from the tank to the fuel pump part of the carburetor and the short pipe directs fuel from the tank cup to the carburetor's mixing chamber.

There are two basic types of throttle mechanisms you may encounter on Briggs engines. Both are depicted in near by drawings. Illustration number 4/24a shows the *cast throttle* and 4/24b shows how to remove it from the carburetor for inspection or cleaning. Illustration number 4/25 does likewise for the *stamped throttle.*

Remove the needle valve for inspection and cleaning. If it is dirty or gummed up, also remove the needle valve seat (ref. no. 147) for a more thorough cleaning. When cleaning carburetor parts, ideally you should use a proprietary engine cleaning solvent made specifically for this purpose. As an alternative, however, a fine grade of kerosene or fuel oil will do the job. Gasoline is too dangerous.

Nylon and rubber carburetor parts should *not* be soaked in engine cleaning solvents, however, or they will deteriorate rather rapidly. A bit of kerosene will have to suffice for cleaning those parts. Replacement, however, is not a bad idea. A new carburetor diaphragm, springs, and nylon pipes together cost less than a package of meat for the family dinner table.

Replace the cleaned old needle valve and needle valve seat or the parts, carefully brushed free of packing materials. Replace the throttle by reversing the instructions which tell how to remove it. Put the choke valve, choke shaft and choke return spring back into place. Reinstall the nylon fuel pipes, and then install the carburetor on top of the fuel tank in the manner already described. Mount the air cleaner, using a new gasket (ref. no. 163) and tighten down the air cleaner mounting screw (ref. no. 534).

Adjusting the carburetor setting must be done after the engine is back in its proper home on top of your lawn mower. If you removed or even moved the carburetor needle valve, chances are good the engine will not start. Turn the needle valve in (clockwise) as far as it will go by hand and without twisting too hard.

Then turn it back out again 1-1/2 turns counterclockwise. Start the engine and allow it to warm up to normal operating temperature, which should take at least five minutes, maybe ten.

Move the engine control to its normal operating speed. Turn the needle valve in (clockwise) until you hear the engine begin to lose speed. The loss of speed comes from the very lean mixture you've just created. Now turn the needle valve out again, counterclockwise, past the point of smoothest operation and keep turning until the engine just barely starts to sound uneven. The uneven operation comes from the very rich mixture this adjustment has created. Now adjust the needle valve to a point almost midway between the too-lean and the too-rich settings. Be a bit partial to a carburetor setting inclined toward the rich side.

Turn the engine control to "SLOW." Adjust the idle speed by moving the idle adjusting screw. On idle, the engine should continue to perform smoothly but at a speed approximately one-half that of the "FAST" setting. This speed will seem very fast to you compared with the very slow idle speed on a six or eight cylinder auto engine. The relatively high idle speed on an air cooled engine with one cylinder is necessary to maintain an adequate degree of lubrication and cooling.

To test your choice of idle settings, move the engine control from "SLOW" to "FAST." The engine should accelerate smoothly and be up to its full governed speed within about two seconds. If the engine stalls, dies out or accelerates too slowly, gradually increase the idle speed setting until the machine meets this acceleration test. In case you should own a tachometer designed for testing a one-cylinder engine, the recommended idle speed is 1750 r.p.m.

CARBURETORS ON BRIGGS & STRATTON ENGINES OTHER THAN THE SERIES 92900 MODELS vary in detail but not often in principle. Before beginning repairs, have on hand a set of gaskets or a complete carburetor repair kit.

On engines larger than the 3.5 h.p. 92900 series which feature a Pulsa-Jet carburetor, the carburetor still mounts on top of the fuel tank but the fuel pump functions are contained not in a machined portion of the tank top but in a fuel pump housing attached to the side of the carburetor. The fundamental function and repair is very similar to what has already been discussed. Illustration number 4/26 shows the fundamental construction of this type of carburetor. Briggs engines which use a Pulsa-Jet carburetor include the model 100902, a 4 h.p. engine common on better 21-inch rotary mowers, and model 130902, a 5 h.p. engine

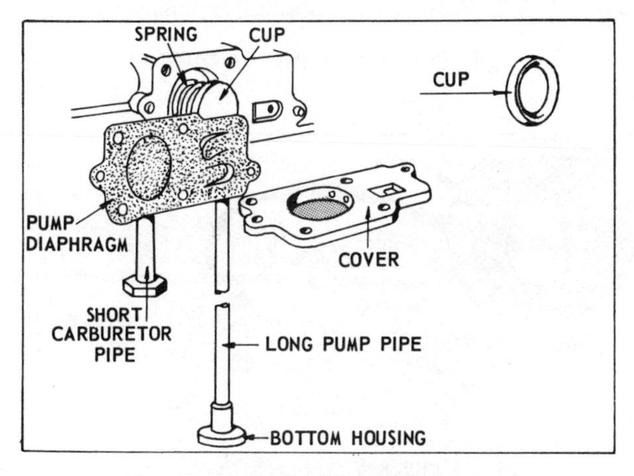

Illustration 4/26. The Pulsa-Jet Carburetor.

found on large self-propelled rotary mowers or small riding mowers.

A 2 h.p. Briggs & Stratton engine such as the Model 60102 is typically found on small reel type mowers. It includes a carburetor which Briggs calls the "Vacu-Jet" and it too mounts atop the gas tank. A manual choke or a "Choke-a-Matic" takes care of its simple choking needs, mainly during starting. Illustrations 4/27 and 4/28 point out significant differences between the Vacu-Jet and carburetors discussed thus far.

Big engines by Briggs & Stratton, from 6 h.p. on up to 8 and

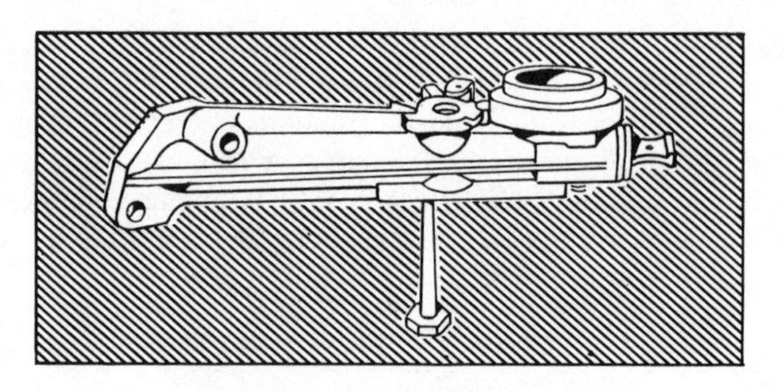

Illustration 4/27. Vacu-Jet Carburetor.

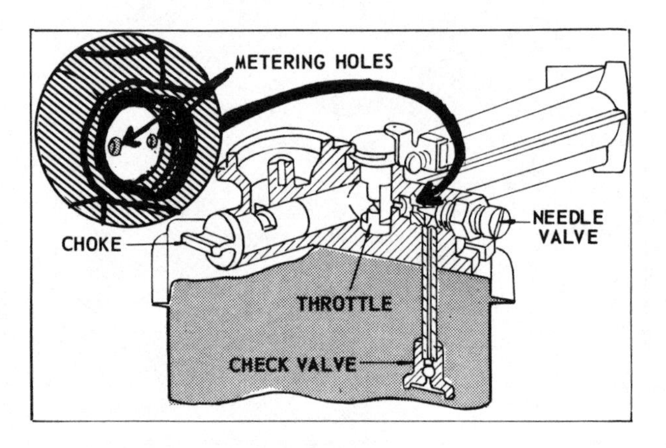

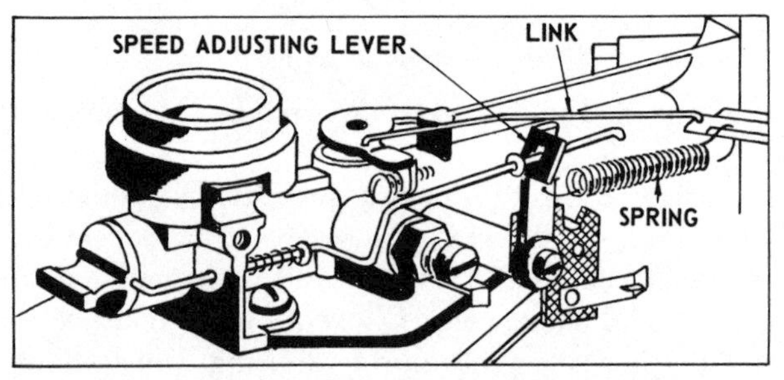

Illustration 4/28. Internal Workings of Vacu-Jet Carburetor.

beyond, include a somewhat more complicated carburetor, the Flo-Jet, which in many ways resembles the workings of carburetors generally found on automobiles. Earlier carburetors discussed *pumped* gasoline out of the fuel tank. In the Flo-Jet system, the fuel tank is mounted higher than the carburetor and gasoline *flows* into the carburetor by the pull of gravity.

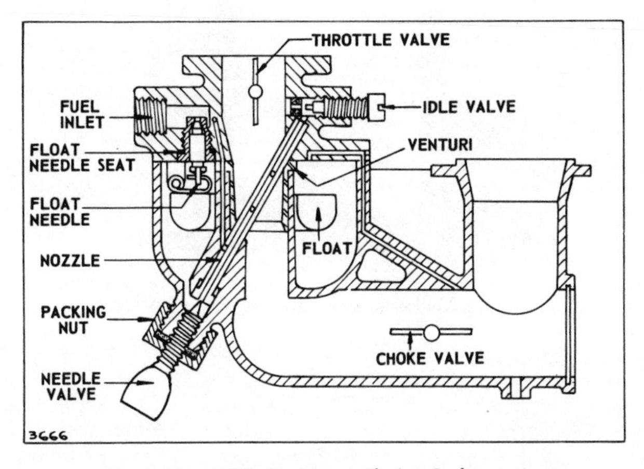

Illustration 4/29. Two Piece Flo-Jet Carburetor.

A float inside the Flo-Jet carburetor regulates the flow of gasoline from the tank in very much the same fashion that the float inside your toilet tank regulates the flow of water into the reservoir tank. Illustration number 4/29 provides a blueprint view of the Flo-Jet carburetor. The float inside the apparatus is round and only in one piece. (A cut-away view in drawing number 4/29 might lead you to believe that two separate floats are incorporated into the carburetor.)

Gasoline in the float chamber is sucked into the nozzle area via the needle valve which again regulates how much gasoline can

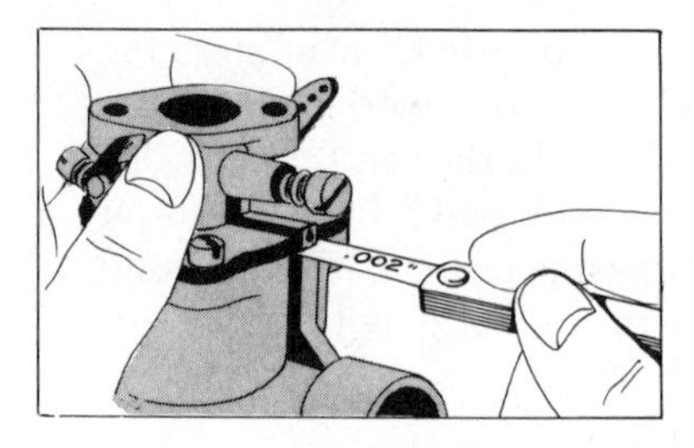

Illustration 4/30. Checking Carburetor Body.

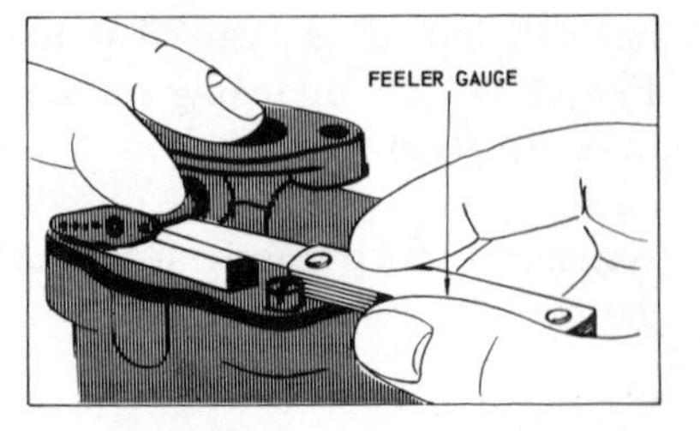

Illustration 4/31. Checking Throttle Wear.

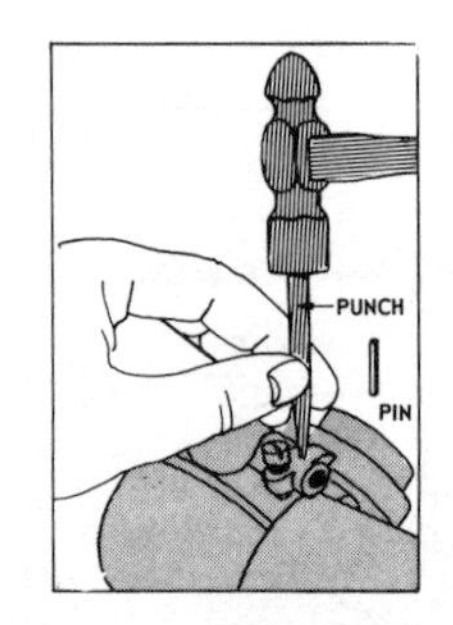

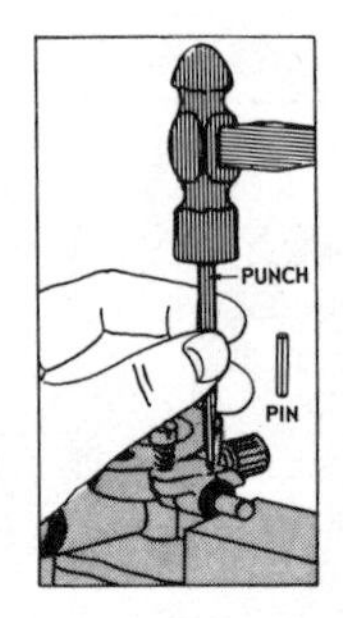

Illustration 4/32. Remove Throttle Shaft and Bushings.

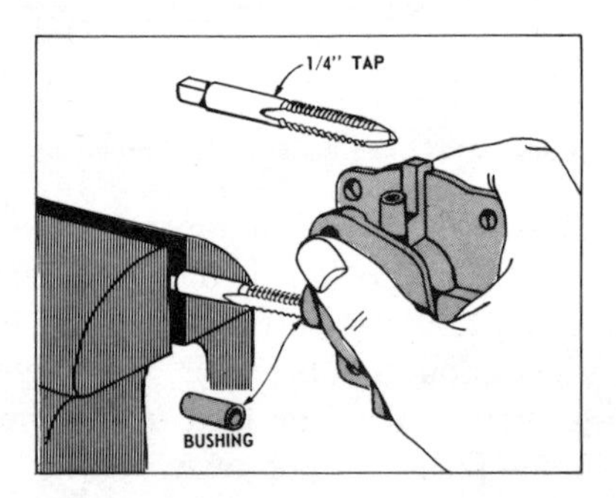

Illustration 4/33. Replacing Throttle Shaft Bushings.

get vaporized into the air stream. Suction comes from the intake stroke of the piston. There are but two discharge holes in the carburetor through which gasoline can be sucked—and the holes are tiny ones at that. When taking apart the Flo-Jet carburetor, these holes must be kept meticulously clean. If they become dirty, they must be cleaned. Do not use wires or drills to clean carburetor holes because they can noticeably increase the size of the openings. In some carburetors, one discharge hole is deliberately made larger than the other; don't try to correct the situation because that's how it was designed to be.

Before disassembling the carburetor, attempt to insert a .002″ feelers gauge between the upper body and the lower body by the air vent just beneath the idle valve. (See illustration number 4/30.) If you can insert such a gauge, the upper carburetor body probably is warped. You can try installing a new gasket (ref. no. 102) and running the feeler gauge test a second time. But if the gauge still slips between the upper and lower body, the upper body must be replaced (ref. no. 91).

To insure smooth engine performance and to avoid throttle jamming, the throttle shaft (ref. no. 97) and the throttle shaft bushing (ref. no. 93), should fit each other tightly enough to prevent wobbling. If you are going to replace the upper carburetor body anyway, the problem of a wobbly throttle shaft and bushing is purely an academic one—the new body includes a new bushing and a new shaft anyway.

You can check for wear in the throttle shaft and bushings by inserting a small, rigid piece of metal or hard plastic between the lip on the carburetor body and the throttle shaft. (See drawing number 4/31.) Hold down the shaft with your finger. It should not quite touch the metal bar. Now measure the space between shaft and bar with a feelers gauge. With your finger *away*, test the space a second time. If the difference between the two readings is greater than .010″, the shaft must be removed from the bushings. If the shaft is worn, replacing only that part might very well cure the difficulty, but don't count on it. If the new shaft does not eliminate the excessive wobble, new bushings will be necessary also.

You may have to spend more for tools to install the new bushings, than the price of a new throttle shaft and bushings. This depends on what your tool box now holds or how well you improvise. Use a thin punch to drive out the one pin which holds the throttle shaft to the throttle stop (ref. no. 100), remove the throttle valve (ref. no. 95 and 96), and then pull out the old shaft. Reverse the order for installing your new shaft.

The old throttle shaft bushings are chewed out with a 1/4″ × 20 tap but with some skill you might be able to substitute a 1/4″ drill. New bushings are pressed into place with a vice—preferably—or judicious use of a vice grip pliers. If the throttle shaft does not slip easily into the new bushings, or if it does not twist easily once in place, ream out the new bushings with a 7/32″ drill.

The float part of your carburetor must float essentially level if it is to control properly the flow of gasoline between the tank and the nozzle. Before reattaching the upper carburetor body to the lower body, turn the upper upside down. It is this part of the carburetor to which the float is attached via a tiny hinge called a *tang*. With a straight edge laid gently across the float, the distance between the carburetor body and the straight edge should be identical on all sides, as demonstrated in drawing number 4/34. If the float happens not to come out parallel to the carburetor body, carefully bend the tang until such time as the float does assume a parallel attitude.

Now you can totally reassemble the carburetor, using only new gaskets all around. When the carburetor is fastened to the engine again and the engine is back on the lawn mower, then you can begin adjusting the carburetor in the same manner as already detailed above for slightly smaller engines. The more sophisticated Flo-Jet carburetor, however, has an adjustment not found on the Vacu-Jet or Pulsa-Jet carburetors, an *idle valve*.

The idle valve establishes the amount of gasoline which mixes with the air while the engine is idling. The traditional needle valve handles that chore while the engine is above idle speeds. Once the idle speed has been set (see drawing number 4/35) the idle valve should be adjusted for the smoothest possible idle.

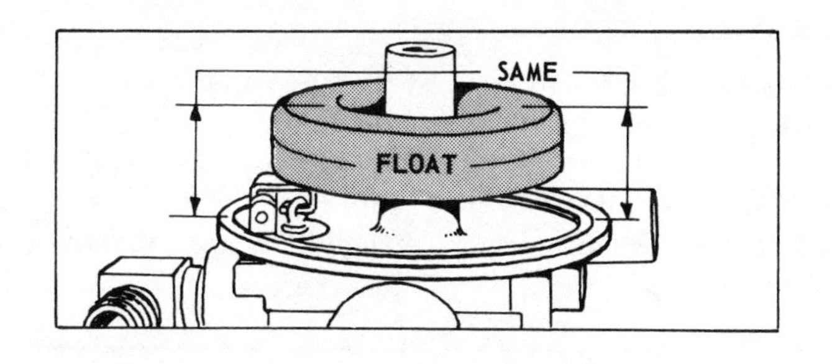

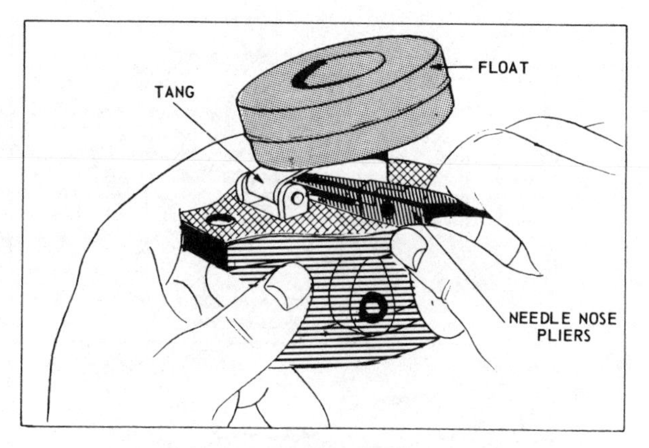

Illustration 4/34. Checking Float Level.

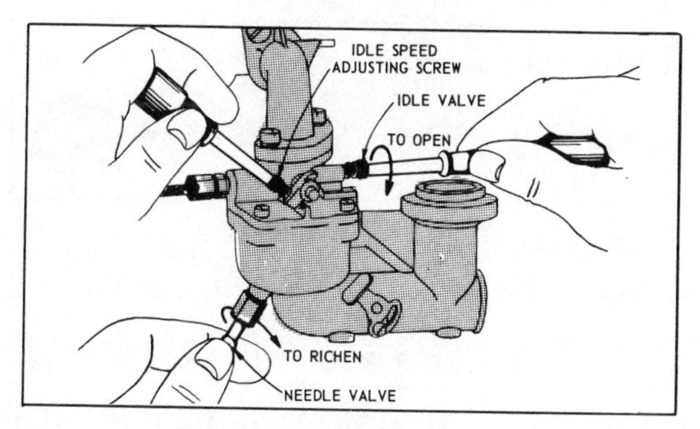

Illustration 4/35. Adjust Carburetor.

It may be necessary to readjust the idle speed upward or downward as the proper setting for idle mixture is being determined.

On the Flo-Jet carburetor, if the engine does not accelerate smoothly from idle, you first should try to solve the problem by turning the needle valve to a slightly richer mixture. Do not

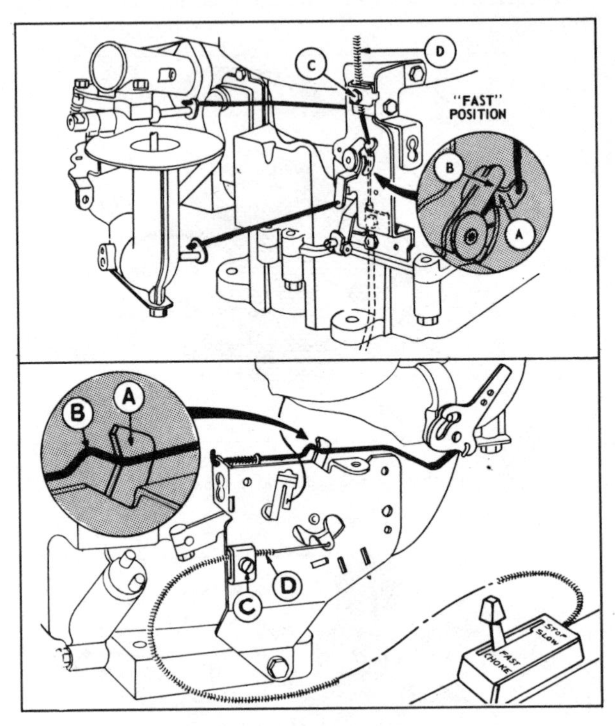

Illustration 4/36. Choke-A-Matic Controls (Typical).

readjust the idle speed as you would on carburetors for smaller engines.

CONTROLS ON LAWN MOWERS REQUIRING LARGE ENGINES, the riding mowers or large self-propelled rotary mowers, come in as many different variations as Carter has Little

Liver Pills. Those with remote controls must have a link between the control levers and the engine choke levers. Very, very few lawn mowers with larger engines seem to employ automatic chokes. On the Briggs & Stratton larger engines, "Choke-a-Matic" is typically employed.

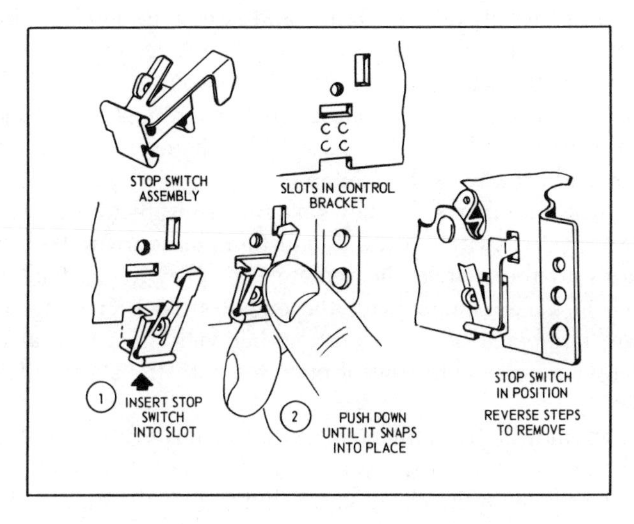

Illustration 4/37 and 4/38. Typical Stop Switch Installation.

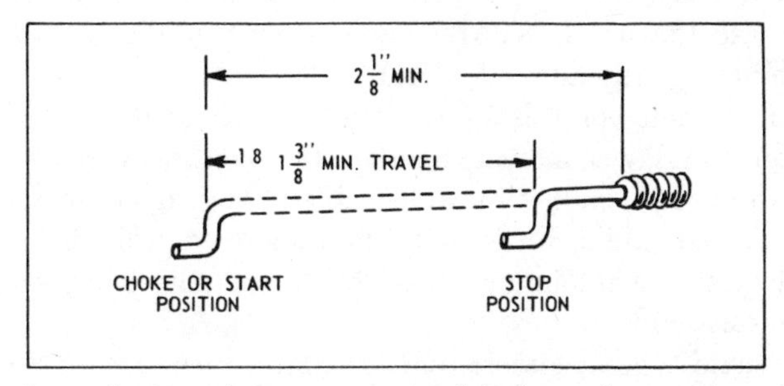

To adjust the usual remote control link to the engine choke, move the lever to its "FAST" position. Consult drawing number 4/36 to understand the A-B-C instructions which follow. The choke actuating lever, "A" in the drawing, should just barely touch the choke link, "B" in the drawing. If not, loosen screw "C" slightly so you can move the casing and wire "D" in or out until lever "A" does just make contact with lever "B". Then retighten screw "C".

At the opposite end of the control lever's movement is often a "STOP" position which activates a switch to short out the ignition, thus bringing the engine to a gradual halt. Such a setting *does not* generally act as a brake and is therefore of only limited value in case of an emergency. As an analogy, you might think of the "STOP" setting as comparable to your turning off the ignition switch on your car; although the ignition stops functioning, the car is still free to coast with whatever momentum the engine and car body still maintain.

Drawing number 4/37 shows a typical installation of a Briggs & Stratton stop switch. Not only must an ignition wire be fastened securely to the switch, the remote control wire must allow the switch to make contact with the engine's block or metal frame. Figure number 4/38 gives yet another view of a typical installation where a remote control operates both the stop switch and choke.

Lawn mower assemblers often erroneously assume that the ideal starting position is always with the engine fully choked, in other words, with the choke valve shutting off the supply of air to the carburetor, thus making as rich a mixture as possible. Consequently, when they label their remote control panels, they generally use the word "START" at one end of the control slide, which in truth should read "CHOKE."

If the remote controls are some distance from the engine itself, say on the lawn mower handle, the control wire generally runs through a hollow, flexible cable. For proper operation of both a choke and a stop switch, the remote control wire must be free to travel a minimum of 1-3/8″ free of the flexible conduit which surrounds it.

A second type of almost remote control incorporates a dial on the blower housing placed there often only to jazz up the lawn mower's appearance. Such an arrangement is particularly dangerous on self-propelled lawn mowers because it demands that you place your arms and legs very close to moving parts to operate the controls.

An engine-mounted dial control may need adjusting after a

certain amount of wear. Figures 4/39a and 4/39b depict a typical layout for such controls. With the switch in the "START" position, loosen wire control screw "A" and move lever "C" until the choke is fully closed. There must always be at least 1/8″ gap

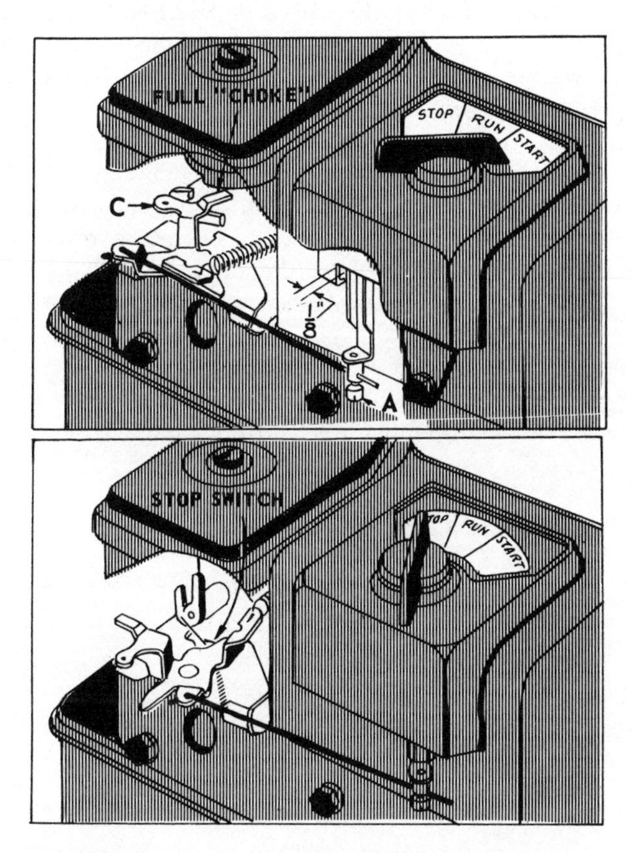

Illustration 4/39. Choke-A-Matic Dial Control Adjustments.

between the control lever and any internal brackets, as illustrated in drawing 4/39a. As shown in drawing number 4/39b, when the dial has been twisted to its "STOP" position, the stop switch should be activated. If the stop switch and choke lever cannot both be reconciled within the limits of your mower's dial and control wire, you must gently twist or bend the choke lever "C" to resolve the paradox as much as possible. For the sake of safety, if you must compromise at one end of the dial, be sure that the "STOP" setting is correct.

GOVERNORS ARE VITAL FOR YOUR SAFETY. On a lawn mower engine, the governor prevents the engine from making too many revolutions per minute which might lead to unstable blade performance in a rotary mower. Excessive speed would definitely make every foreign object struck by the blade into more of a lethal weapon than under normal operating conditions.

The maximum safe r.p.m. figure for a rotary lawn mower depends upon the blade size. In order to operate a rotary lawn mower with even relative safety, the tip of the blade should never travel faster than 19,000 feet per minute; the shorter the blade length, the faster the engine will rotate at the center to build up 19,000 feet per minute at the tip of the blade.

From the Briggs & Stratton manual, the following blade lengths reach 19,000 feet per minute at the corresponding engine r.p.m.'s.

BLADE LENGTH	ROTATIONAL R.P.M.
18″	4032
19	3820
20	3629
21	3456
22	3299
23	3155
24	3024
25	2903
26	2791

Briggs & Stratton recommends that the governed speed of any lawn mower engine should be set 200 r.p.m. lower than the maximum speeds shown.

Unless you own a tachometer, one which can be used with a moving shaft, do not tackle major adjustments to the governor on your rotary lawn mower. Reel mowers are another matter altogether. Keep all exposed parts of the governor clean and avoid bending or distorting the various arms and linkages connected to the governor on your engine. Admittedly, small engines seldom are capable of going much faster than the 3400 r.p.m. which would theoretically be entirely safe with a 20″ or even 21″ blade. However, at 7 h.p., a Briggs engine can spin at 4000 r.p.m. which would be fully 30% too fast for a 24″ blade.

The purpose of a governor on a gasoline engine actually is not only to insure safe operation—it is there to insure smooth and consistent operation under varying loads. If you run a mower across a patch of neatly trimmed grass, the engine purrs along at a cool, even pace. Move into some weeds or heavy, long grass, however, and the increased load would normally slow down the engine. A governor counteracts the effect of increased load by opening the throttle further so the engine can generate enough power to maintain an almost constant speed despite the heavier demand upon it.

On most Briggs & Stratton engines below 5 h.p., an air vane governor is employed. The flywheel, as we already know, acts as a blower in addition to all of its other duties. The wind blown across the engine by the fan blades not only cools the engine, it applies pressure to the air vane of a governor. The governor spring tends to hold the throttle open and the governor vane tends to hold it closed. Thus, if the engine speeds up, more wind is blown onto the vane which in turn cuts back on the throttle.

Larger engines now generally employ a mechanical governor. Small, movable, counterweights are spun indirectly by the crankshaft. As the crankshaft spins faster, the weights are moved further and further apart by their own centrifugal force. The outward motion of the weights partially closes the throttle and brings the engine back to its governed setting.

INSIDE THE BRIGGS & STRATTON ENGINE is one cylinder, one piston, two valves and a small collection of gears. They needn't overwhelm a novice fix-it man or woman. In the first place, a lawn mower engine has but one set of pistons and valves to keep timed, compared to six or eight in your family car. In the second place, Briggs & Stratton has built an image for itself

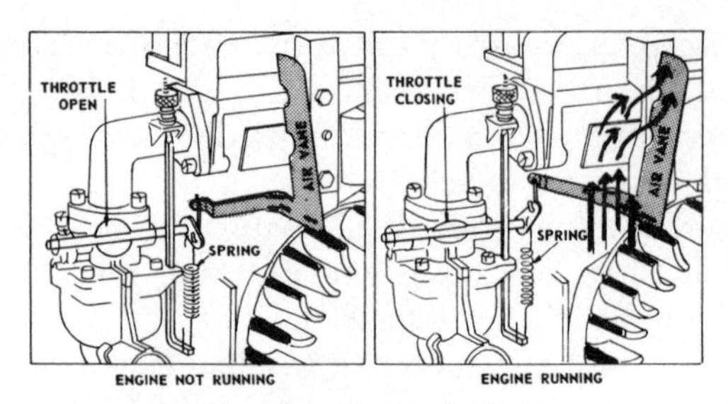

Air Vane Governor

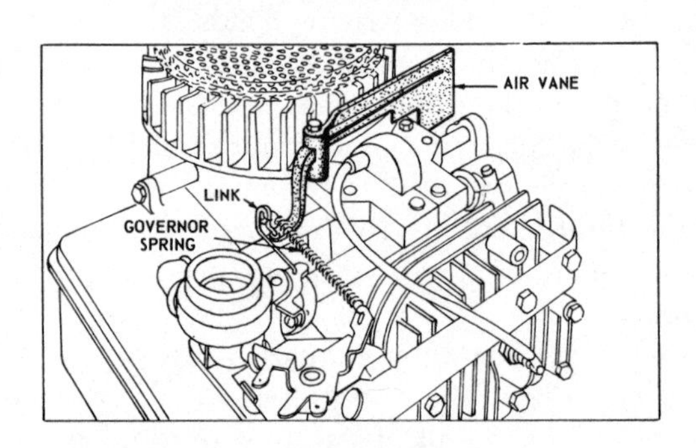

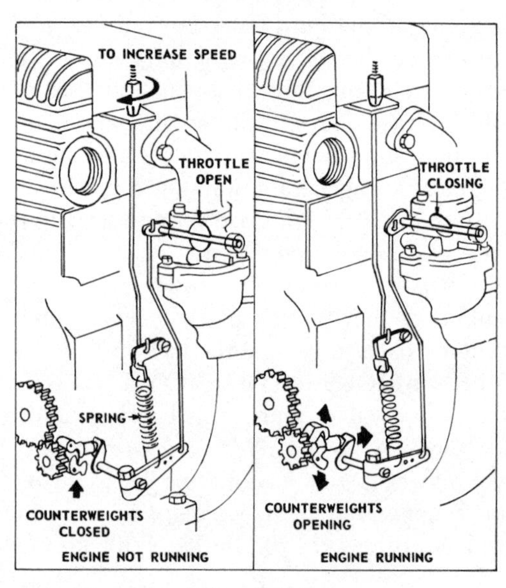

Illustration 4/40 and 4/41. Governors.

as the manufacturer of repairable engines. This is an image which lawn mower salesmen seem to enjoy obscuring for the sake of a faster sales turnover. And third, you can avoid repair problem caused by your lack of specialized repair equipment by *not repairing* valves and related paraphernalia—*replace* them! A valve grinding and refacing job on your lawn mower engine could run close to five times the price of a complete new set of valves, springs, retainers and valve seats for a model 92000 series engine, assuming you install the new ones yourself.

We've jumped the gun, however. Do you really need a new valve job? To find out, consider these symptoms.

Are you adding oil to the crankcase several times a year but oil seals and gaskets in the block don't leak the oil onto the floor of your garage? Does Old Betsy seem to have lost her horse power, but really lost it and not just suffered emotional shock at the sight of a shiny new mower on the neighbor's lawn? How is your compression? Grab the flywheel and spin it counterclockwise—that's backwards. If it rebounds smartly, chances are your engine's vital parts are in good shape and you should look externally for the problem. It's easy enough to find out for sure.

Taking wrench in hand, expose that mysterious world down under by removing the head bolts in the cylinder (ref. no. 5), but remove them systematically. Start twisting them loose only half a turn at a time and in the sequence shown in drawing number 4/42. This is to prevent warping the aluminum cylinder head. In replacing the head, the same procedure must be followed.

Heat and gasoline and oil generally will have combined to seal the gasket and the engine block and the cylinder head into one unmovable chunk. A few cautious probes with a screwdriver, perhaps even a firm tap or two against the end of the screwdriver, and the head should come off the block easily. Discard what's left of the gasket.

Next remove the oil pan, which Briggs & Stratton prefers to call an oil *sump* (ref. no. 18). As the name implies, it's full of oil which you will want to drain out via the drain plug (ref. no. 15). Since every part you touch during this part of the repair operation will be oily and greasy, you should clean each part in a good and proper solvent as you remove it.

As might be expected, Briggs & Stratton offers a tool specifically made for removing the valves with ease, a #19063 valve spring compressor. Various tool suppliers offer similar inexpensive devices. But judicious and enlightened use of a screw driver can

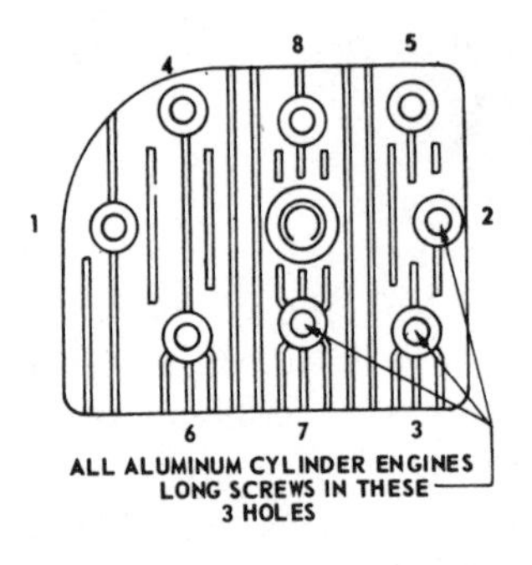

Illustration 4/42. Order in Which Head Bolts Must be Removed or Replaced on Block of Engines. It is very important that the head of your engine's block be kept from warping when bolts are removed or replaced. Consequently, engine manufacturers specify the exact order in which they should be tightened. Either tighten or untighten them gradually, but always in this order.

handle the removal almost as well. The valve spring has to be squeezed together to allow for removing the retainer, a trick illustrated in drawing number 4/43. With the retainer gone, the valve slips upward and out with relative ease.

Through the cavity exposed when you removed the oil sum, locate the connecting rod (ref. no. 29) and remove it. The two bolts which hold the connecting rod to the crankshaft are protected against vibrating loose by the connecting rod lock (ref. no. 31). Bend the two tabbed ends of the lock out of the way with a hammer and screwdriver or small chisel. Unscrew the two bolts and the connecting rod cap opens neatly to release the connecting rod from the crankshaft. At this point, the crankshaft is hanging with only the upper bearing (flywheel end) for support. You must either remove the crankshaft altogether—by first pulling the flywheel—or exercise great caution that it does not get bent. A new crankshaft can be expensive.

Before sliding the piston out of the cylinder, remove carbon deposits from the lip of the cylinder, a treatment you should also apply to the valve seats and anywhere else that shows an accumulation of that annoying stuff. Heavy deposits are scraped loose first with a knife and then with a wire brush. If you do not remove the bead of encrusted carbon which often accumulates at the top of the cylinder, the piston rings can be damaged as

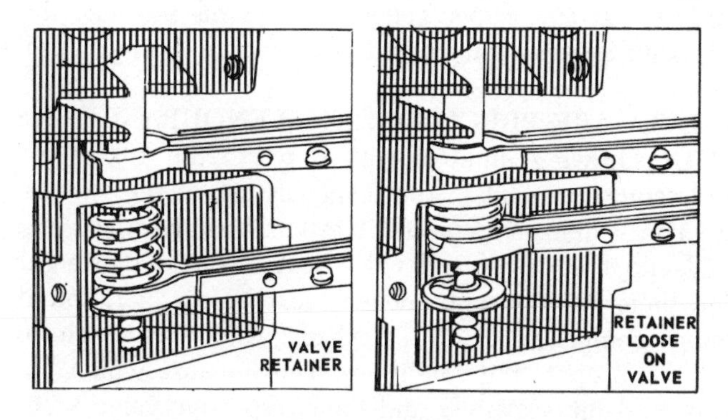

Illustration 4/43. Removing Retainer and Spring.

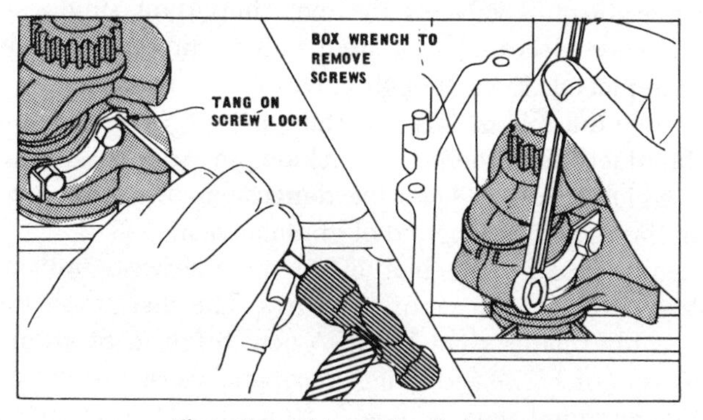

Illustration 4/44. Bending Rod Lock.

you slide the piston out. There is even a special tool just for removing carbon from the cylinder lip should you feel prosperous.

A home lawn mower used properly just a few times every month during the warm months is unlikely to wear out a set of piston rings for at least several years. Likewise for the valves. There are exceptions, however, like an engine over-powered by heavy turf and weeds, or an engine which has to run at steady high speeds continuously, even an engine mounted on a piece of equipment which really ought to have been given a more powerful engine in the first place. And with the kind of persistent use a professional landscaper generally gives his mowers, they can go through a set of valves and rings within a single, profitable season.

VALVES ARE PRECISION, HIGH-ENDURANCE INSTRUMENTS. Drawing number 4/45 is worth the 10,000 words it would require to explain the nomenclature of the various parts of a valve system. Do consult it. After the valves have been removed and the valves and valve seats cleaned thoroughly, examine them for signs of uneven wear or discoloration due to burning. Temperatures on an exhaust valve can reach 5000°F. under tough operating conditions. The pressure which both the exhaust and intake valves and their respective valve seats must be able to withstand can reach 500 pounds per square inch. Even slight blemishes can severely decrease an engine's power.

The margin of a valve is the most important single area to measure when deciding if a used valve can be reinstalled or must be replaced. Even though valves and valve seats are made of extremely durable metals, a certain amount of wear is inevitable. Manufacturers, therefore, include an area known as the margin which serves to keep the dimensions of a face relatively uniform throughout the period of gradual wear.

Carefully measure the margin on an otherwise undamaged, clean valve you might continue to use. The margin should not be noticeably smaller than 1/32″. A new Briggs & Stratton valve has a margin of 1/32″ and Briggs recommends discarding a valve when its margin becomes 1/64″. However, instruments like a micrometer capable of measuring accurately to 1/64″ are not likely to be laying around a kitchen junk drawer or a basement work room.

For the valve seat, your measurement should fall within the range of 3/64″ to 5/64″, or in other words, no more than 1/64″ above or below 1/16″ wide. Since temperatures at the exhaust valve are higher than at the intake valve, the exhaust valve seat

VALVE PART NAMES

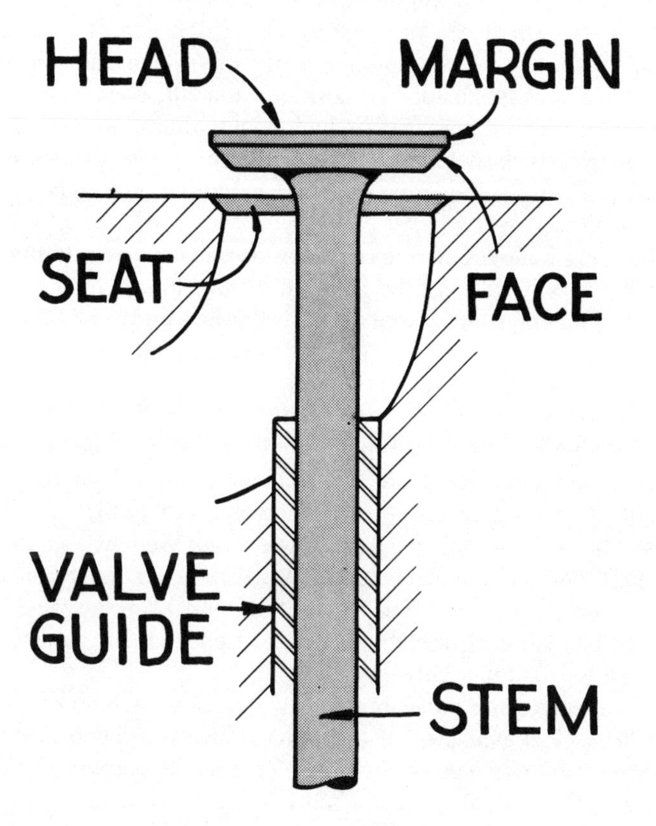

Illustration 4/45. Nomenclature for Parts of a Valve.

is the one most likely to require attention. The tools and mechanical expertise needed to replace a valve seat insert are probably somewhat beyond a novice handyman. It is a procedure not often called for, fortunately, and thus it can best be left for authorized factory-trained repairmen.

Once you have a new valve (and perhaps a new seat), or clean and usable ones, they must be *lapped* into place, which is mechanics' jargon for polishing them smooth. Most good hardware stores or auto supply stores sell lapping compound. You will need a small can of coarse and a small can of fine. Smear some of the coarse compound on the valve face and seat when the valve is in its customary position. A small wooden stick with a suction cup at one end—which looks almost like your son's toy arrow—is used to rotate the valve quickly with firm downward pressure so the valve face and valve seat are both polished by the effort.

Lapping removes all fine scratches or pits which could diminish a valve's ability to hold the 500 pounds per square inch pressure. Move from the coarse compound to the fine and when the valve parts are just as shiny as a new mirror, meticulously clean away every trace of the lapping compound.

Set the valves into their seat, sliding them through the valve guides and down to the tappets which ride on those two tough little cams of the camshaft. These cams raise each of the valves about 25 times a second. Turn the crankshaft until one valve is raised to its uppermost position. Then rotate the flywheel by hand through one complete revolution. With a feelers gauge, measure the clearance between tappet and the end of your valve stem. Check it against figures given below. Repeat this measuring sequence for the other valve.

If the valves you measure are new ones, you will no doubt find too *little* clearance. And if old valves have too much clearance, they must be replaced. Adjust the clearance by carefully grinding or filing the end of the stem until it falls within the prescribed limits. If too little clearance exists between the tappet and the valve stem, after heat expands both metal parts, the valves might not close fully. The result would be both loss of compression and badly burnt valves which would very soon need replacing once more.

If you tolerate too much clearance between tappet and valve stem, valves will not open soon enough and will close too soon. The result will be loss of power and inefficient engine operation.

	INTAKE VALVE		EXHAUST VALVE	
	max.	min.	max.	min.
ENGINES WITH ALUMINUM CYLINDER*	.007	.005	.011	.009
ENGINES WITH IRON CYLINDER	.009	.007	.019	.017

* Virtually all modern Briggs & Stratton engines of 7 h.p. and under have aluminum cylinders which go by the trade-name of "Kool-Bore."

Not all engines use identical exhaust and intake valve springs so make sure if they do differ, that the stronger one goes with the exhaust valve. Oil all moving parts before reinstalling and then reverse the instructions for removing valves when reassembling your engine.

If your engine is subjected to constant use or persistent abuse, or if valve wear is a problem for some other reason, you should consider using two available products which will reward you with longer life for your valves. A *stellite exhaust valve* costs more because it is made of a tougher alloy than common valves. But its additional cost is more than made up for by its longer life. And a *valve rotator,* used with either a common valve or the stellite valve, helps to overcome one of the problems leading to shortened valve life. When a piece of carbon lands on the valve seat or valve face, it prevents the valve from closing fully and eventually causes the valve face to become burnt prematurely. A valve rotator gives the valve a slight spin on each lift stroke, thus tending to wipe away deposits which might otherwise remain. The valve rotator, or "Rotocap" as Briggs & Stratton labels its product, slips very simply into place between the valve spring and valve stem retainers with no special tools. Drawing number 4/46 shows the "Rotocap" installation.

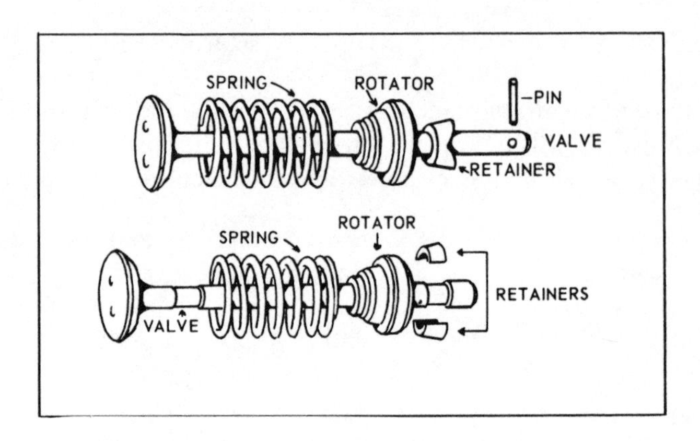

Illustration 4/46. Stellite Valve and Rotocap.

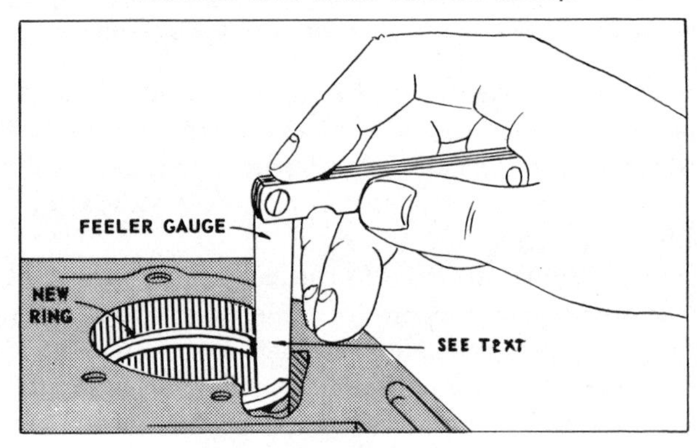

Illustration 4/47. Checking Ring Gap.

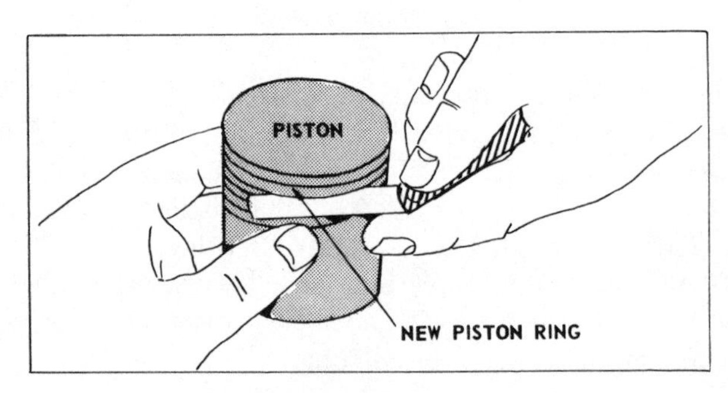

Illustration 4/48. Checking Ring Grooves.

A NEW SET OF PISTON RINGS for a model 92900 series Briggs & Stratton engine is not a major investment if you install it yourself. However, if you carefully remove each of the three rings from the piston—assuming you have some reason to suspect they are causing trouble—you can test the rings. Be careful to note in what order they belong because each is a bit different from its mate.

Insert the rings one at a time about one inch down into the cylinder. The ring should fit snugly against the cylinder all around. There will be a gap in the ring, which is very essential because the rings expand from the heat of your engine. Measure the gap with a feelers gauge and reject the ring if a gap is greater than .035″ for the compression ring, the top one on the cylinder, or .045″ for the other two. (These reject figures are for aluminum engines. On a cast iron engine the compression ring gap is .030″ and for the others, .035″.)

Before installing new rings, clean the carbon accumulation out of the grooves. The end of an old ring often works very nicely for scraping away carbon. You'll also find oil return holes in the ring grooves; these must be cleaned well. New rings are inserted in their proper order working from the crankshaft end of the piston toward the top. Freely lubricate the rings and piston with a good grade of motor oil and clamp a *piston ring compressor* over the new rings. There is very little way to improvise a makeshift ring compressor and very little need to either since it is so inexpensive.

Figure number 4/48 shows a test of the piston which should be made after *new* rings are on and after the rings have been clamped once by the piston ring compressor. With the ring compressor removed, a feelers gauge measures the space between ring and piston wall groove. If a .007″ gauge can be inserted, the piston itself is worn. Briggs recommends that a worn piston be replaced. This is enough of a major undertaking so that you might want to spend a winter evening reading a Briggs & Stratton repair manual to learn how to test for a worn cylinder wall and what to do about it.

Figure number 4/49 shows how the piston is started into the

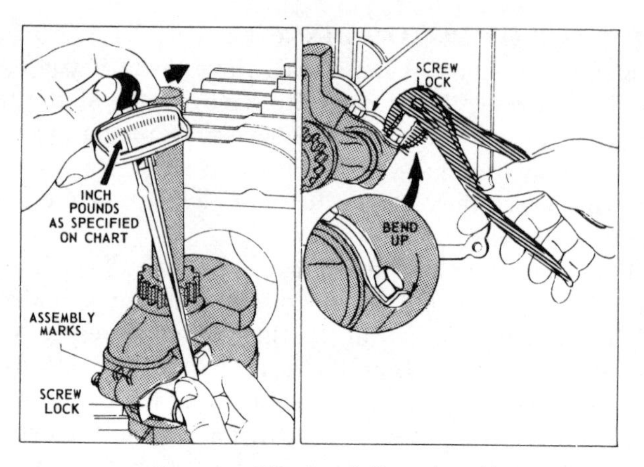

Illustration 4/49. Install Piston Assembly.

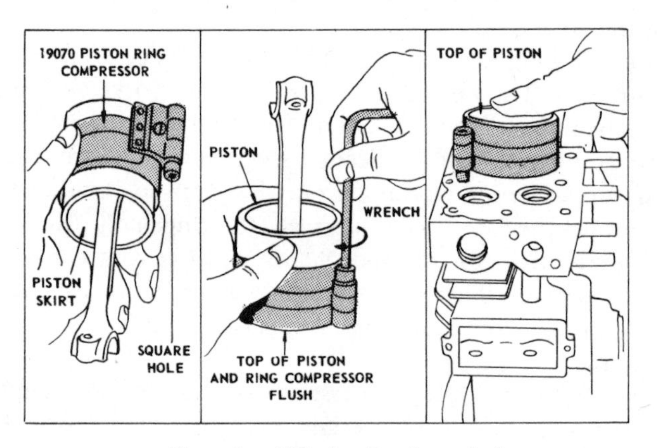

Illustration 4/50. Bending Screw Locks.

cylinder with the ring compressor in place. Without loosening the compressor any more than is absolutely necessary, shove the piston downward into the cylinder with a soft tool such as the *wooden handle* of a hammer. The ring compressor must remain in place; otherwise the piston ring's natural expansion would prevent your reinstalling the piston and rings.

The piston rod can be fastened to the crankshaft at this point. Be sure the surface of all moving parts is thoroughly oiled first.

The connecting rod assembly must go back together in exactly the same way that you took it apart. Some of the assemblies are shaped so that their correct position will be obvious. If not, look for what Briggs & Stratton calls *assembly marks.* They generally are just enscribed lines one on each side of the two parts where the connecting rod assembles around the crankshaft. Drawing number 4/50 shows how to tighten the two bolts with a torque wrench and then bend up the screw locks with a pliers. For en-

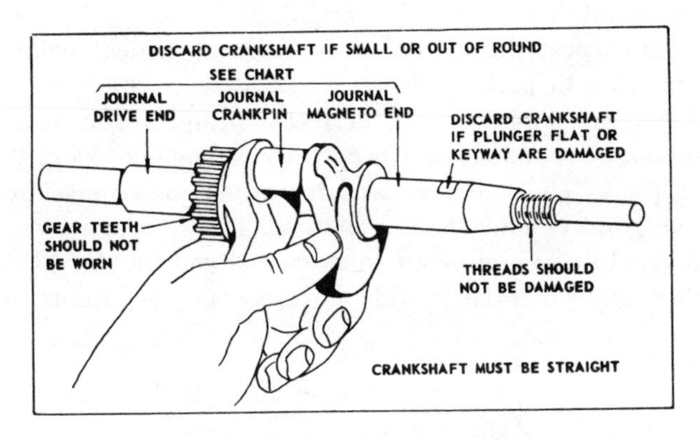

Illustration 4/51. Crankshaft Check Points.

gines up through the model 130000 series, Briggs recommends 100 inch-pounds (8 foot-pounds). For the model 140000 series and up, 140 inch-pounds (12 foot-pounds) is recommended. If you don't have a torque wrench, tighten the two nuts very firmly but make a real effort to tighten both nuts equally. Before bending the screw locks, rotate the engine by hand through two or three cycles to make certain that the moving parts are moving properly and do not strike any part they are not supposed to strike.

If you ever have to remove the crankshaft from your engine, first remove any dirt or rust or even metal burrs from the power take-off end, that is the end fastened to the mower blade for instance. The sump is held in place by several bolts and then by a gasket which quite often is just as sticky as the cylinder head gasket. Before reaching in to pull the crankshaft out, first rotate the shaft until its timing mark lines up with a corresponding marking on the cam gear. If you fail to perform this small ritual first, you could easily spend hours of tedious trial and error in trying to time the valves and pistons and magneto correctly when you're ready to bolt the engine back together. Finally, lift out the cam gear and then remove the crankshaft.

If you inspect the crankshaft visually at critical points, you may be able to find points of excessive wear. First of all, the crankshaft must be straight, perfectly straight. The gear teeth should not be rounded and worn; badly damaged or worn threads at the power end will only lead to further mechanical troubles at some point in your engine's lifetime. The keyway which holds the magneto and crankshaft in proper alignment should not be rounded, gouged or damaged in any other serious fashion.

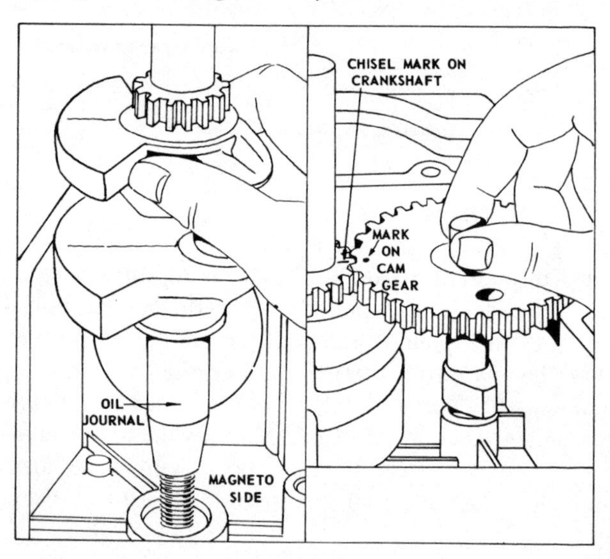

Illustration 4/52. Aligning Timing Marks.

Wear caused by bearings is hard to assess without a micrometer unless it has become excessive. Figure number 4/51 shows the various parts of a crankshaft which can lead to vibration or overloading if wear has reduced excessively the size of the shaft. A micrometer is a relatively expensive instrument designed to measure with an accuracy of .0001″. Briggs assumes its mechanics have one on hand when deciding the fate of a suspect crankshaft. Their figures (below) which recommend at what point a crankshaft should be considered worn out, will be useful if you own a micrometer. Otherwise they may offer merely a numerical guide to supplement your instincts. When new, the crankshaft on a model 92000 series engine is 7/8″ or .875″.

	CRANKSHAFT REJECT SIZE (IN INCHES)		CAM REJECT SIZE	
BASIC MODEL SERIES ALUMINUM CYLINDER ("KOOL-BORE")	DRIVE END	MAGNETO END	CRANK-PIN	CAM LOBE
6B, 60000	.8726	.8726	.8697	.883
8B, 80000* 820000, 92000*	.8726	.8726	.9963	.883
100000, 130000	.9976	.8726	.9963	.950
140000, 170000	1.1790	.9975†	1.0900	.977
190000	1.1790	.9975†	1.1219	.977
CAST IRON CYLINDER				
14, 19, 190000	1.1790	1.1790	.9964	1.115
200000	1.1790	1.1790	1.1219	1.115
23, 230000	1.3759	1.3759	1.1844	1.115

* Auxiliary drive models P.T.O. bearing reject size: 1.003″ and cam lobe .751″.

† Synchro Balanced Magneto Bearing reject size: 1.179″.

Reverse the removal order to reinstall the crankshaft and cam gear. Illustration number 4/52 will help. The crankshaft, when it is properly installed, must be free to slide back and forth in the bearings; mechanics call this *end play*. Install a pulley or "blade protector" or blade adapter firmly to the power end of the crankshaft. Just before tightening the device, push on the crankshaft in the direction of the magneto and maintain that pressure until all screws in the pulley or other device are tightened fully. Now shove the crankshaft from the other end and measure with a feelers gauge how much end play exists. There should be .002–.008″ of end play.

If *too little* end play is found, an extra gasket must be added to the sump. The gasket which normally is used between the oil sump and the main engine block is .015″ thick. Briggs & Stratton has available extra gaskets of .005″, .009″ and .015″ to deal with end play problems. Various combinations of gaskets should be tried until the end play falls safely within the established limits of .002 to .008″.

If *too much* end play is the problem, Briggs offers a thrust washer which slips onto the power take off end of the crankshaft. The washer significantly reduces end play and must generally be used in combination with one or more of the three supplemental gaskets. Under no circumstances should you end up with less than .015″ of gaskets for the sump.

INSIDE THE SUMP IS AN OIL SLINGER which does exactly that—it slings oil around inside the sump area to keep moving parts well oiled. Excess oil drips back down into the sump to be picked up by the slinger again. This system of lubrication is simplicity itself. Although there is seldom anything which befalls the slinger, check it every time the sump is pulled and should it ever start showing signs of wear, replace the inexpensive little gismo. It is held very simply in place against the cam gear by a spring washer which slips over the cam gear shaft.

At each end of the crankshaft is a leather or hard rubber oil seal (ref. no. 3 and 20) which provides insurance against oil slipping out of the engine and into the magneto or onto the lawn mower blade. It would be peny-wise for non-pros not to replace them whenever the crankshaft has been removed or replaced. Briggs & Stratton mechanics seem to enjoy a certain status when they can successfully reinstall a crankshaft and not need new oil seals. New oil seals must first be smeared liberally

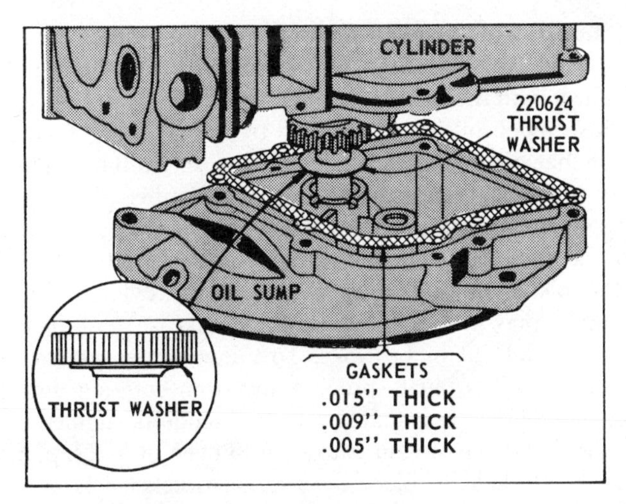

Illustration 4/53. Correcting Crankshaft End Play.

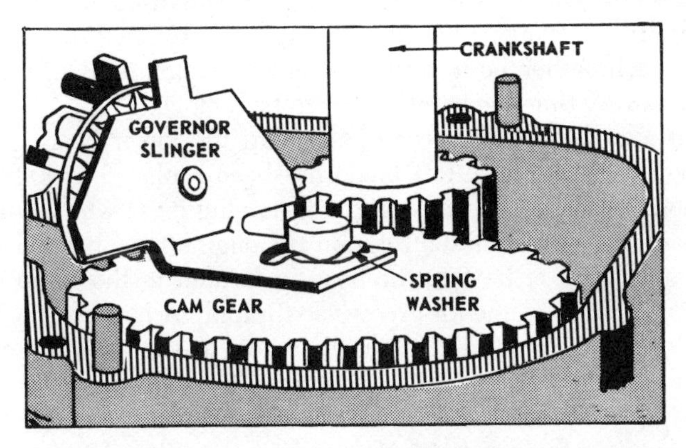

Illustration 4/54. Oil Slinger and Bracket, Vertical Crankshaft Engines.

with "Lubriplate" or similar silicone-based white grease and then pressed into place, sharp edge toward the engine, until the seal is flush with the hub.

With so much oil being slung so freely at every moving part, some of that oil would be likely to squeeze by the piston rings, breaker plunger, oil seals and many, many other places . . . *except* that engines are designed to operate with a vacuum inside the crankcase to prevent all of those possible oil leaks. Whenever the piston makes a downstroke in the cylinder, it would build up a pressure inside a closed crankcase and *that* really would force oil out of bearings. To allow air to get *out* during piston downstrokes but to prevent any from entering during upstrokes, a *breather* is installed on most modern engines ranging in size from the one in your car down to even little 2 h.p. models.

A fiber disc held in place by a metal bracket acts as a valve which allows air to escape but not reenter the crankcase. Assuming the breather on your engine looks OK physically, you can check the disc's efficiency with a .040″ gauge. Briggs & Stratton recommends a wire gauge such as you used to gap spark plugs. If the gauge can enter the space between the fiber disc and the breather body, install a new breather. And use a new gasket every time the breather has to be removed.

On rotary lawn mowers in particular, the lower bearing takes a good deal of abuse. If a blade has been replaced or balanced, if the crankshaft appears straight, if moving parts which come in contact with the crankshaft within the engine have been checked out and found to be OK, the bearing located in the sump could very well be to blame for excessive vibration. A bad bearing there can lead to eventual difficulties elsewhere as the vibration causes other parts to wear out faster than normal.

Briggs & Stratton offers an extremely inexpensive gauge which gives quite a foolproof test of a bearing condition. You can order one for your specific engine model. Unfortunately, such a gauge is not the sort of thing the typical tinkerer thinks of ordering in advance, so your instinct and two or three of your senses may have to be your guide. The bearing and crankshaft should fit so snugly, as you replace the sump cover, that if the cover is tilted so much as 1/8″, you will not be able to slide the cover easily back in place over the crankshaft.

Briggs makes no provision for replacing the lower bearing on its smaller engines commonly used by lawn mower assemblers. The proper technique for dealing with a worn bearing in engines from the 92900 series and down is to replace the entire sump, not an expensive alternative at all. It probably would prove more economical for you to replace the sump as a bearing repair tactic even on those engines which *do* feature replaceable bearings.

The chore of properly replacing an engine bearing requires specialized tools which do not readily lend themselves to improvised substitutes. The job of replacing the upper bearing in a lawn mower engine can best be turned over to a qualified mechanic with proper equipment.

You've just breezed through the rudiments of small gasoline engine repairs—some pretty heavy rudiments at times. If you feel like getting deeper into engine repairing, out of necessity, as a hobby or out of professional interest, your next step would be to invest just a few dollars in the Briggs & Stratton parts and service data. That manual includes an exploded parts break down for every current engine in their line, a price list for replacement parts and a fairly comprehensive general repair guide which will prove especially useful for tackling heavy Briggs & Stratton repairs. The techniques carry over to competitors' engines quite well. And the book is a bargain.

If just the one engine from your own lawn mower interests you, you can save a little change by buying separately the general repair instructions and the exploded parts diagram for your particular engine model number.

What Betty Crocker did for your kitchen, Briggs & Stratton has done for your garage.

CHAPTER 5

Engine Repair II—Other Engines and Starters

BRIGGS & Stratton does not make the only engines found on lawn mowers. Some of the other manufacturers you are likely to find include Clinton, Tecumseh-Lauson and Kohler. Each of them maintains a network of authorized service stations to handle repairs and to stock replacement parts. In many localities, the same agency represents all or most of the small engine manufacturers.

Small engines, up to about 3.5 h.p., are generally Clinton or Tecumseh if they aren't Briggs & Stratton. At the upper end of the spectrum, the 7.5, 8, 10 and even 12 h.p., engines not made by Briggs are typically Kohler or Tecumseh. Not every engine manufacturer is as free and easy with his repair manuals as Briggs & Stratton. Often citing matters like warrantees, some of them prefer that owners of lawn mowers using their engines take the ailing machine to an authorized repair station. The engine warrantee is generally good for only one year. What then?

Some of the lawn mower and small tractor suppliers have incorporated engine repair techniques and specifications in the service manuals for their lawn mower. The Gravely Corporation includes Kohler engine repairing in its manuals and a bit about Onan engines. John Deere & Company features comprehensive Tecumseh and Kohler engine repairs. Some local factory authorized repair stations are willing to order engine repair manuals for a customer but most direct you to the national office of each engine manufacturer.

It will be of considerable help for you to know *which* manufacturer's engine has been included in your particular brand of lawn mower. Some assemblers seem to make an effort at concealing this knowledge from a buyer. The *Briggs & Stratton* identification code already has been discussed in the previous chapter. *Kohler* engines have a "K" in front of the model number and then three digits generally follow. The three digits represent an engine's displacement (the size of its cylinder, in other words) times 10. The Kohler K241 engine, for instance, has a cylinder displacement of about 24 cubic inches, which produces approximately 10 h.p. The K301 has about a 30 cubic inch displacement, or 12 h.p. K321 turns out 14 h.p. And a K181 generates 8 h.p. with its approximately 18 cubic inch displacement.

Tecumseh engines are sent from the factory with a letter and number code also. The letter or letters identify whether the engine has a horizontal or vertical crankshaft. The numbers which follow represent the horsepower times 10. The Tecumseh V50 engine, therefore, is a vertical crankshaft engine having 5 h.p. If your mower does not specify the engine manufacturer but you come across a plate on the engine or an entry in the parts list which reads "HH100," you'll know that you own a Tecumseh horizontal crankshaft engine with 10 h.p. The Tecumseh LAV35 is another engine commonly found on some lawn mowers, a vertical crankshaft engine with a power rating of 3.5 h.p.

Clinton engines are identified with a 10 digit number code without any preceding letters. Briggs & Stratton likewise uses an all number code but it never extends beyond 6 digits. Clinton's code inevitably begins with a "4" or a "5". Briggs has no "4" or "5" in its lead-off digit at the moment.

Clinton, like Briggs, identifies engineering features in its 10 digit code. The 10 digits, incidentally, are broken up into a set of 3, then a set of 4 and a final set of 3 digits. Digit number one will almost always be "4" on lawn mowers, a 4-cycle engine. Clinton also makes the simpler 2-cycle engine to which it gives a code "5". Digits number 2 and 3 identify basic engine series; an odd number here is used for vertical crankshaft engines and an even number on horizontal versions.

The fourth digit in a Clinton code identifies the starter: 0, recoil; 1, rope; 2, impulse; 3, crank; 4, 12 volt electric; 5, 12 volt starter-generator; 6, 110 volt electric; 7, 12 volt generator.

Bearings are identified in the fifth digit by Clinton: 0, standard bearing; 1, aluminum or bronze sleeve with flange and pilot used for mounting auxiliary equipment on the engines crankshaft; 2, ball or roller bearings; 3, ball or roller bearings with flange and pilot. Digits number 7, 8, 9 and 10 are factory design and model codes.

At the end of the *serial number*—not the 10 digit code number—there is often a letter code which identifies part changes which occurred during the run of a particular engine type. An engine with a "C" letter code might look and operate identically with one sporting a "D", but some vital part would not be interchangeable between the two engines.

A typical Clinton code likely to be found on lawn mower engines is "405 0000 070." The "4" tells you it is a 4-cycle engine. The "05" is for a vertical crankshaft engine, the type most commonly found on rotary mowers. In the fourth position, the "0" indicates a recoil starter and the following "0" (fifth position) tells you the engine has an ordinary type of bearing. The sixth position "0" lets you know that the engine had no facilities for auxiliary power take-off when it left the Clinton factory.

There are three basic Clinton engines used for most lawn mower applications; only the first three digits are needed when referring to basic types. They are the 405 and 415, both 3.5 h.p. engines, and the 4.5 h.p. model 498. The first two have aluminum cylinders and pistons, just about standard for small lawn mower engines by all manufacturers. The Clinton 498, however, is made of cast iron, a heavy but rare and rugged feature to be found on an engine so small as 4.5 h.p.

GASOLINE ENGINES ARE REMARKABLY SIMILAR IN THEIR FUNDAMENTAL MAKE UP no matter who the manufacturer is. You should be able to apply the Briggs & Stratton repair section to other engines. This chapter will dwell mainly on principal differences in ignition systems and carburetors, two of the most frequently serviced items. It will also cover starters for all engines including Briggs & Stratton. All of the engine manufacturers include starters which are so identical they might very well all have come off the same assembly line.

AIR CLEANERS on the larger Tecumseh and Kohler engines are of the *dry element type*. They should not be washed. Gentle tapping is the recommended cleaning method. Elements should be replaced after every 100 hours of use or whenever they become really dirty. For engines which operate in very dusty conditions, a polyurethane band pre-cleaner is available. The pre-cleaner slips over the air filter cleaner inside the housing and can be washed occasionally. Heavy blasts of compressed air can rip a dry element air cleaner internally, rendering it useless. If one proves to be quite dirty, replace it.

Many novices, and some repairmen too, take a rather casual attitude toward the lowly air cleaner since it doesn't seem to be a hefty, imposing, moving part of the engine. If dirt slips through the cleaner element or through a faulty gasket, however, it will more than likely scratch pistons, piston rings, valves or other moving parts and seriously hamper their proper performance. An executive at Clinton Engines feels that only one lawn mower owner in a thousand bothers to clean his air filter properly. The Clinton executive feels that neglected air cleaners are one of the major contributors to shortened engine life.

Gasoline on some large engines flows through a sediment bowl which gives heavy foreign matter space to settle out of the gasoline before being sucked or pumped into the tiny orifices of the carburetor or into the rapidly moving parts of the engine itself. Most engines which feature the sedimènt bowl have a shut-off valve on the gas tank itself. To clean the sediment bowl, shut off the fuel tank valve and then loosen the thumb nut until the bowl itself can be slid out. Empty it, wash it and then dry thoroughly before putting it back into place. Use a new gasket whenever the bowl is removed for cleaning. You ought to keep one or two of them on hand for emergencies.

After the sediment bowl is reassembled but *before* the fuel shut-off valve at the tank is reopened, disconnect the fuel line at the carburetor. Now open the shut-off valve and wait until

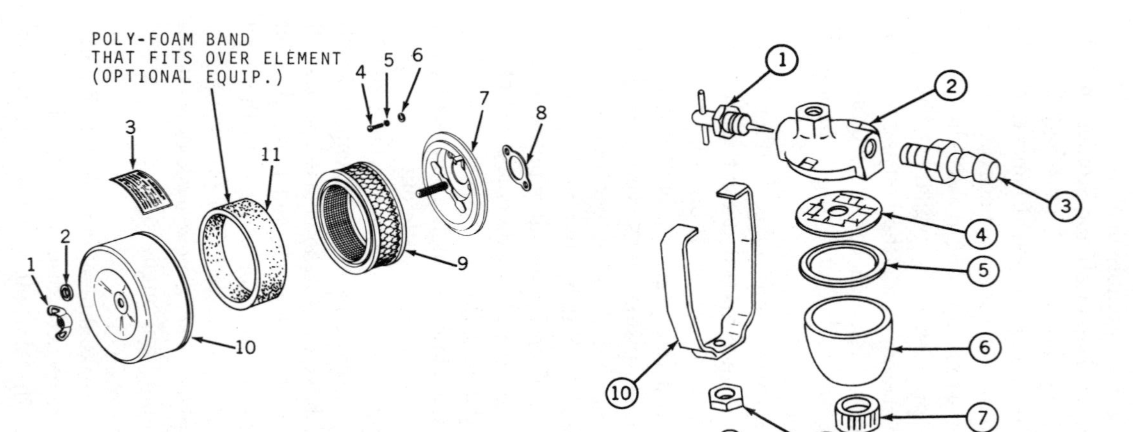

1—Wing Nut
2—Copper Washer
3—Decal
4—Machine Screw
5—Lock Washer
6—Plain Washer
7—Base
8—Air Cleaner Gasket
9—Air Filter Element
10—Air Filter Cover
11—Polyurethane Pre-Cleaner (Extra Equipment)

Exploded View of Air Cleaner Components

Illustration 5/1.

1-Needle Shut-Off Valve
2-Base
3-Connector
4-Screen
5-Gasket
6-Bowl
7-Thumb Nut
8-Machine Screw
9-Nut
10-Bail

Exploded View of Sediment Bowl

Illustration 5/2.

gasoline begins flowing out the end of the fuel line before reconnecting the line to the carburetor. This allows air to escape safely. Otherwise it could form a bothersome air lock in the carburetor or fuel line.

CARBURETORS ON LARGE ENGINES come in many makes and models. Their function is essentially similar, although their controls and adjustment screws are located in varied spots. The exploded drawings and cutaway views (Illustrations number 5/3, 5/4 and 5/5) printed near by will provide insight into where the important adjustments are located on various carburetors.

Clinton's LMV (also known as LMG and LMB) Carburetor, Kohler's Carter Carburetor and Tecumseh's Walbro Carburetor are broken down and cleaned in solvent. On the Carter, do not attempt to remove the choke and throttle assemblies since the screws holding them in place are peened. If they are damaged, it is less expensive to replace the entire carburetor than to attempt to repair the choke or throttle parts according to John Deere's repair manual. Repair kits are available to replace frequently worn parts for almost all carburetors.

Aside from the mixture and idling settings, the float is about the only major internal adjustment to be made. First make sure that the float is parallel to the machined surface of the top of the carburetor body, as was done with the Briggs & Stratton carburetor, and then adjust the float so it rests at the proper height above the machined body. On the Carter carburetor, gently bend the lip until the float will rest easily on top of a 3/8″ drill resting on the machined body. The Walbro carburetor specifies a 9/64″ clearance; you may use a 1/8″ drill as a close substitute if 9/64″ is unavailable.

Some models of Clinton's LMV carburetor feature a screw which can adjust the float to 5/32″ above the carburetor rim when the carburetor is held in an inverted position. Other models have to be bent near the hinge if a screw adjustment is unavailable.

A SEPARATE FUEL PUMP is found on large Kohler engines whereas almost all smaller engines have a fuel pump system built into the carburetor itself. A very few Briggs & Stratton large

Illustration 5/3A. Clinton LMV-type carburetor.

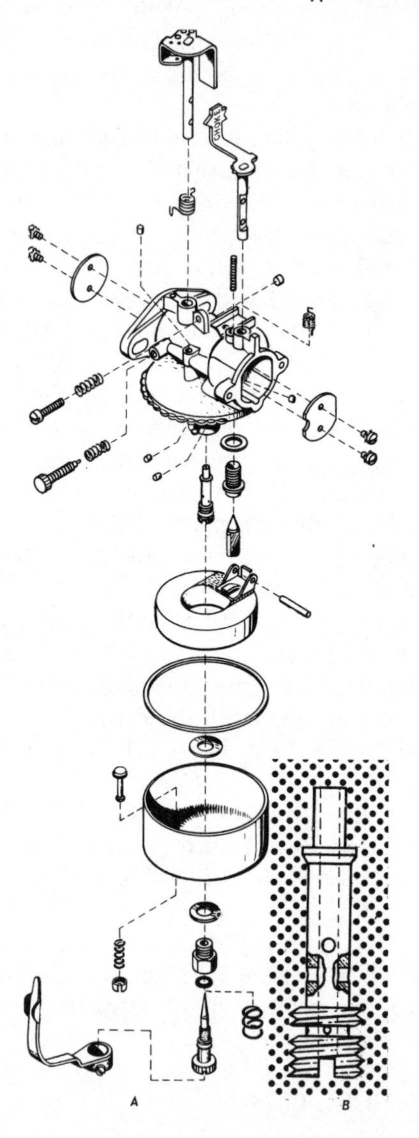

Illustration 5/3B. Power needle adjustment in Clinton LMV type carburetor.

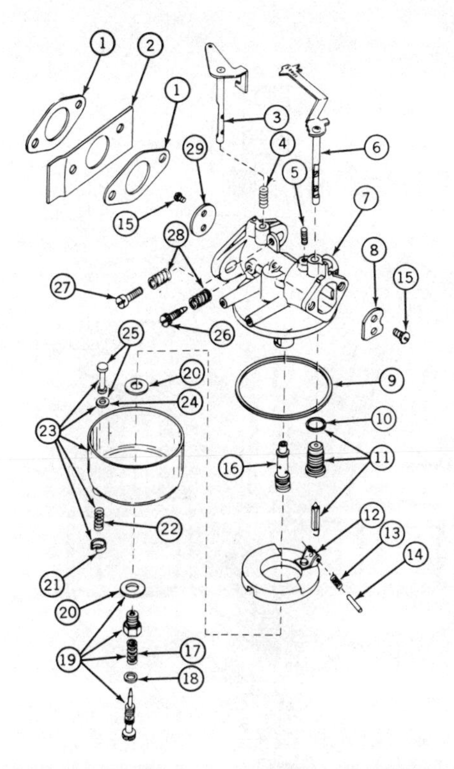

1-Gasket (2 used)
2-Air Baffle
3-Throttle Shaft and Lever
4-Throttle Return Spring
5-Choke Stop Spring
6-Choke Shaft and Lever
7-Carburetor Body
8-Choke Shutter
9-Gasket
10-Seat Gasket
11-Float Valve and Seat Assembly
12-Float
13-Float Spring
14-Float Pin
15-Machine Screw (4 used)
16-Main Nozzle
17-Adjusting Needle Spring
18-Gasket
19-High Speed Adjusting Needle Assembly
20-Gasket
21-Retainer
22-Bowl Drain Spring
23-Bowl and Drain Assembly
24-Drain Stem Gasket
25-Drain Stem
26-Idle Adjusting Needle
27-Idle Adjusting Machine screw
28-Adjusting Needle Spring (2 used)
29-Throttle Shutter

Exploded View of Walbro Carburetor

Illustration 5/4A.

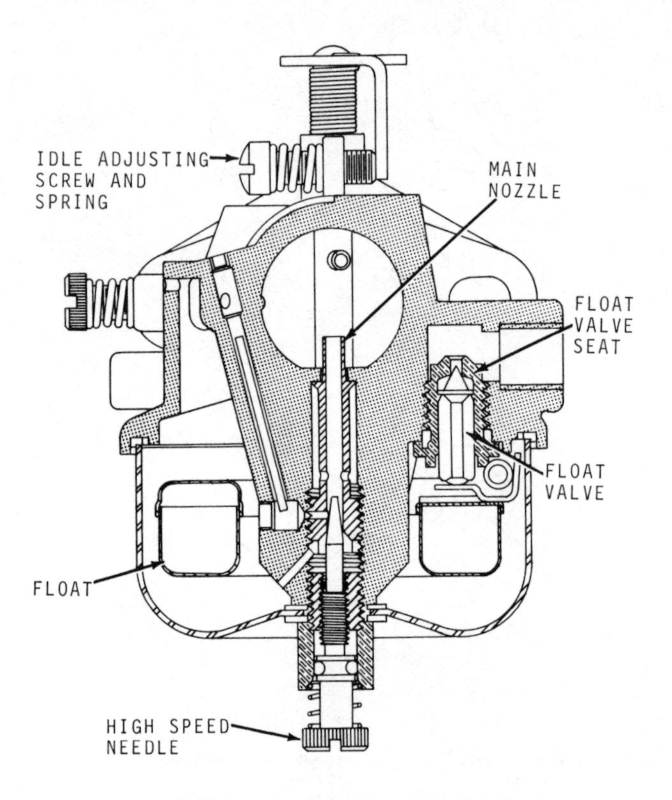

Cutaway View of Walbro Carburetor

Illustration 5/4B.

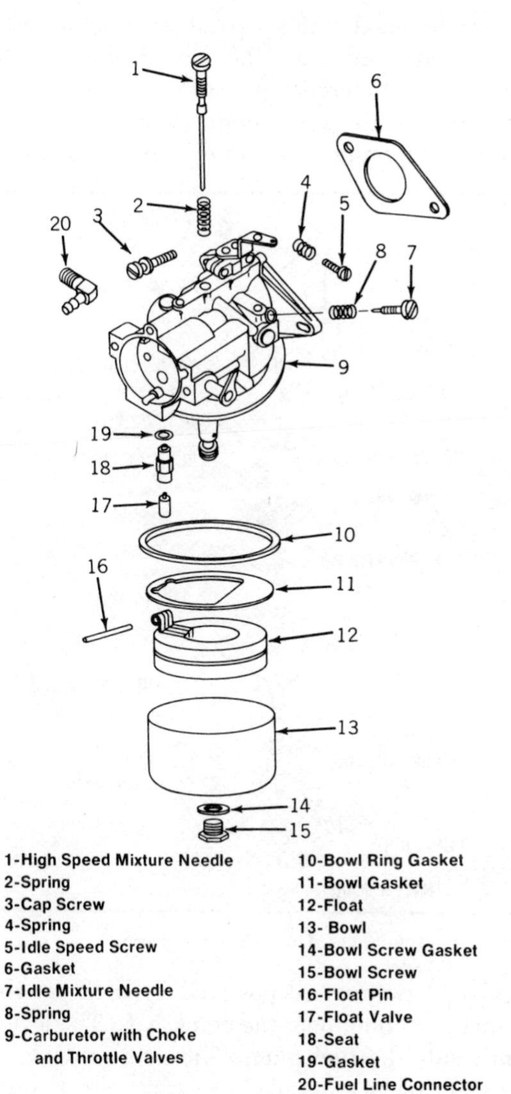

1-High Speed Mixture Needle
2-Spring
3-Cap Screw
4-Spring
5-Idle Speed Screw
6-Gasket
7-Idle Mixture Needle
8-Spring
9-Carburetor with Choke and Throttle Valves
10-Bowl Ring Gasket
11-Bowl Gasket
12-Float
13- Bowl
14-Bowl Screw Gasket
15-Bowl Screw
16-Float Pin
17-Float Valve
18-Seat
19-Gasket
20-Fuel Line Connector

Exploded View of Carter Carburetor Components

Illustration 5/5.

engines also are designed with a separate fuel pump. A faulty fuel pump can be diagnosed with the aid of the troubleshooting chapter (Chapter 3). References to carburetor diaphragms in that chapter apply also to the fuel pump which is really little more than an external diaphragm fuel supply for the engine.

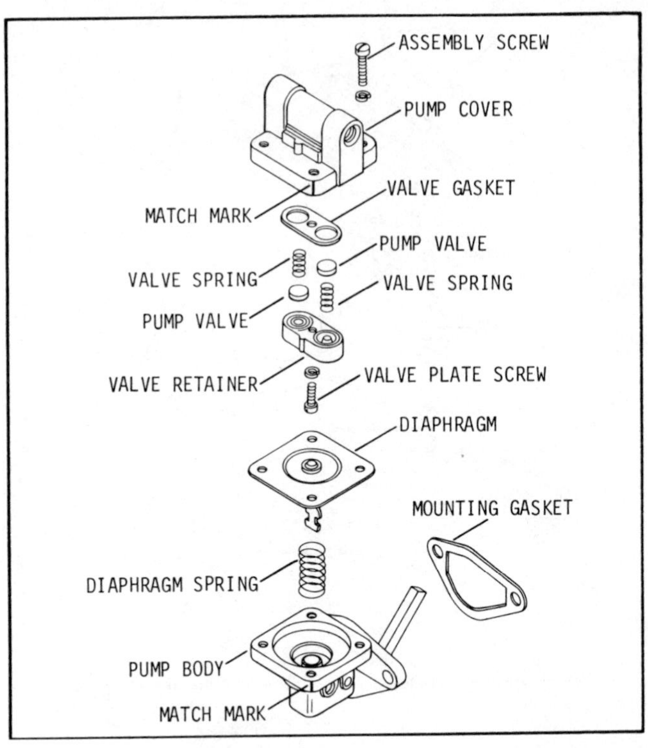

Illustration 5/6. Kohler Fuel Pump.

Mechanical fuel pumps are powered by a lever which rides on an additional cam built onto the crankshaft. The lever actuates a diaphragm inside the fuel pump. Kits are available which replace the moving parts most likely to wear out or break inside the pump, principally the diaphragm, gaskets and sets of springs. Exploded drawing number 5/6 shows the assembly of all parts in a fuel pump used on Kohler engines. The "match mark" is an identifying scratch which *you* should scribe onto the upper and lower parts of the pump body so you can later reassemble the pump with both pieces aligned in the same direction as they had been when you started.

Remove the diaphragm from the fuel pump lower body by depressing the diaphragm bracket and turning it 90° (1/4 turn) to unhook it from the lever. Reverse this technique to reassemble the pump with a new diaphragm. *Do not tighten* the four assembly screws more than is absolutely necessary to hold the pump body together. Then, holding only the pump body, push the actuating lever as far as it can travel. Keeping the lever depressed, tighten all four assembly screws. If you fail to follow this technique, the diaphragm is likely to be stretched.

With a new mounting gasket, bolt the pump back onto the engine, connect the fuel lines, and start the engine. Always check for gasoline leaks after tampering with fuel lines, carburetors or pumps.

IGNITION SYSTEMS IN BIG GASOLINE ENGINES by Kohler and Tecumseh vary greatly, not only in their layout of parts but in some very fundamental theories of design. Clinton's ignition system is very similar to Briggs & Stratton's. The *magneto and breaker* system is about the only ignition you will encounter on all Briggs & Stratton or Clinton engines used in lawn mowers. Kohler and Tecumseh use magnetos and breakers also, but some models employ the *battery and breaker* system, the *breakerless solid state* system and the *motor-generator* system.

Ignition points generally are not located under the flywheel on Kohler or Tecumseh models larger than 5 h.p. Their basic setting should be .020″ as with Briggs & Stratton and Clinton. Clinton gaps its spark plugs at .025″. Spark plug gaps for Kohler and Tecumseh vary according to engine make and model although the most common setting on Kohler engines is .025″. You must consult your own user manual for the correct spark plug and its proper gap.

There is very, very little which can go wrong with a properly maintained magneto system of a good lawn mower engine. But a badly set gap on a spark plug, especially if ignored for too

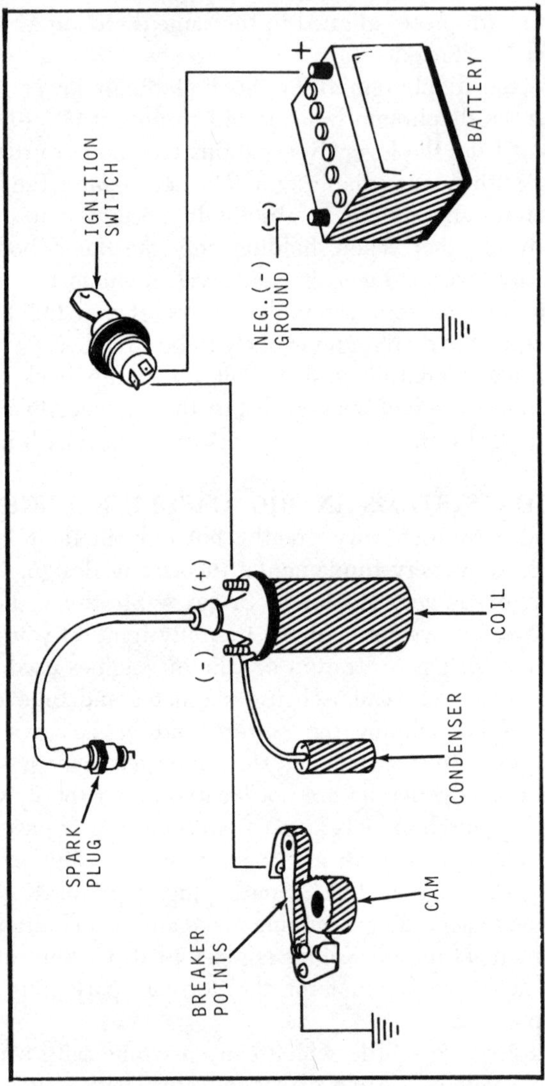

Illustration 5/7.

long, can make such heavy demands on a magneto that it may have to be serviced or replaced.

With the *battery and breaker* ignition system, a lead and acid storage battery similar to the one in your car is used to provide electric power for the spark coil. To keep the battery charged, an alternator spins with the engine and generates alternating current—similar to house current. The alternating current is converted to direct current—similar to battery current—by a rectifier so it can recharge the battery. An alternator is roughly mid-way between a magneto and a generator. The tractor alternator uses permanent magnets, as does the magneto, but delivers a virtually continuous current, as does the generator.

With the exception of the spark plugs and ignition points, there are few other ignition parts in a battery-breaker system which can be serviced without sophisticated electronic test equipment. One exception is a *fuse* located on the rectifier-regulator assembly or within related wires on those systems having one.

Storage batteries should be kept clean; the water level in them should be up to the indicated level at all times, and they must be protected from discharging under freezing conditions. A discharged storage battery can freeze at fairly moderate temperatures.

On a motor-generator system which receives heavy use, the *brushes* should be checked every 200 hours of use. If the carbon in the brushes has worn away 1/2 its length, replace the brushes at your earliest opportunity. The old brushes can be reinserted and used for several more operations, however, but try to put them back in exactly the same position you found them.

Fortunately, many local auto mechanics will be able to help you diagnose and correct ignition difficulties with the large lawn mower and small tractor engines because the components are basically very similar to some automotive ignition systems and the same test equipment can be relied upon. More than likely you will need a repair manual from the engine manufacturer or the lawn mower assembler to provide your auto mechanic with electrical specifications peculiar to the lawn mower components you own.

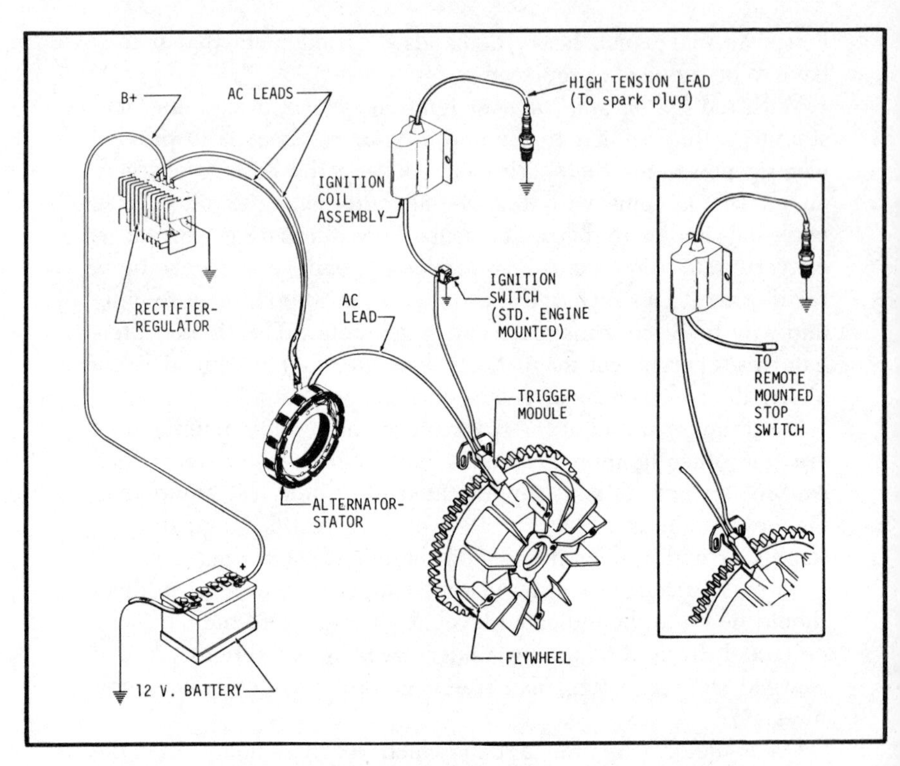

Illustration 5/8. Typical Breakerless Ignition System.

THE BREAKERLESS SOLID STATE IGNITION SYSTEM does away with the need for moving parts like ignition breakers and relays. Transistors and diodes electrically regulate the alternator's output, doing away with relays. And another set of transistors and diodes senses at what instant the spark plug should receive its high voltage charge, thereby eliminating the condenser and ignition points. There are no parts in a breakerless solid state ignition system which can be serviced without sophisticated equipment. Fortunately, such an innovation seldom requires servicing. Kohler says in a repair manual for its engine models featuring the breakerless solid state system, "With the breakerless ignition, timing is permanently set for the lifetime of the engine. Except for the spark plug, the entire system is virtually service-free."

Tecumseh also offers an engine with solid state ignition system. Their manual explains how to take voltage and ohm readings at various terminals, which you can do too if you have proper equipment. Also, mechanics test the device by first removing the ignition switch (to guard against shorts) and then connecting the high voltage lead to a test spark plug that is grounded to the engine. When the starter turns over the engine, you should hear and see a spark jump across the 1/4-inch gap in the test plug. If not, replace the ignition module.

HAND STARTERS ALL SEEM TO LOOK ALIKE but few of them *are* exactly alike. The hand operated ones have basically the same set of parts: a rope which indirectly spins the flywheel either via a gear or a friction clutch, and a spring which rewinds the rope. Alternately, the rope first winds the spring and the spring spins the engine. Some starters are wound up by a small crank and later activated when a lever on the lawn mower handle is released, the impulse starter. There is one thing to keep firmly in mind about all starters which employ springs—approach the spring with great care and then only when absolutely necessary. It is wound under quite a bit of tension and can damage moving parts of your engine or your body if that tension is released carelessly.

The Tecumseh LAV series engines have a recoil starter geared to the flywheel gear. When you pull the starter rope, the big nylon starter gear slides over to mesh with cogs on the flywheel gear and spins the engine, hopefully starting it. Pulling the rope also tenses a spring which will automatically rewind the rope for you.

To disassemble Tecumseh's LAV series starter, pull out the rope slightly and hook it into the V-notch of the mounting bracket. Untie the knot to remove handle and then release the rope. Now you can unscrew the starter cover secure in the

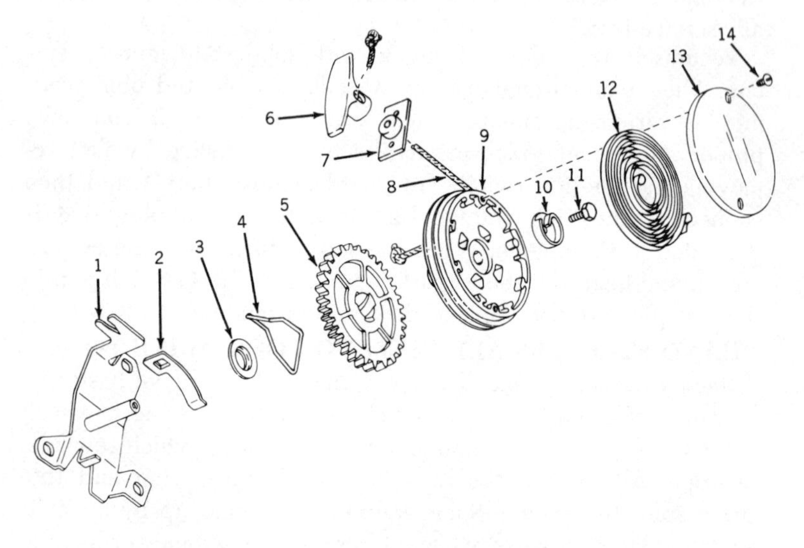

1—Mounting Bracket	*8—Starter Rope*
2—Rope Clip	*9—Pulley Assembly*
3—Thrust Washer	*10—Spring Hub*
4—Brake	*11—Hub Screw*
5—Gear	*12—Rewind Spring*
6—Handle	*13—Spring Cover*
7—Rope Bracket	*14—Screw (2 used)*

Recoil Starter Components for LAV Series Engines

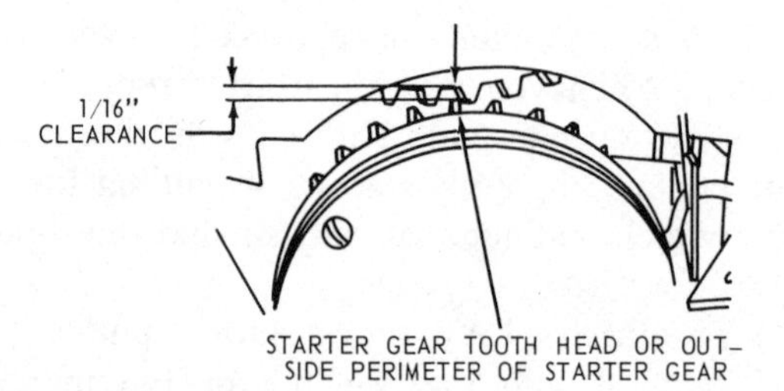

Illustration 5/9 A & B.

knowledge that the spring no longer is tensed. (This assumes that the starter has been working up until the time of your check-up.) If the spring is in good condition, do not remove it from the pulley but do apply a thin film of light grease such as "Lubriplate" to the pulley.

If the rope needs replacing on the LAV series Tecumseh engine, or if parts beneath the spring pulley need replacing or inspection, remove the entire starter from the mounting bracket. Then remove the hub screw and spring hub, and separate the gear from the pulley without removing the spring. *Do not lubricate* the brake groove in the gear; friction at that point forces the starter gear into mesh with the flywheel.

Reassemble parts to the LAV series Tecumseh starter in the numbered order shown on drawing number 5/9. Grease the bracket shaft with "Lubriplate." The two ends of the brake wire straddle the clip parallel to the bracket shaft on the side of the bracket which has two mounting holes. The inside hook of the spring slips into the spring hub and the hub screw is tightened very securely. If you own a torque wrench, the hub screw should be torqued to 50 inch-pounds (4 foot-pounds).

After the LAV series cover is secured in place with its two screws, tighten the spring with 2-1/2 counterclockwise turns. Hook the rope into the V-notch of the bracket while you bolt the starter back onto the engine. There should be 1/16″ clearance between the outside portion of the starter gear and the inside portion of the flywheel gear, as shown in drawing number 5/9b. Put the starter handle back onto the rope with a knot and release the rope.

TECUMSEH'S "H" AND "V" SERIES ENGINES have a somewhat different type of starter than the "LAV" series. Exploded drawing number 5/10 shows the arrangement of parts. Any worn or damaged parts must be replaced, including the rope. With all rope starters, it is important that any new rope be the same length and thickness as the one it replaces.

Initial tension is put on the Tecumseh rewind spring before fully assembling the "H" and "V" starter apparatus. Viewed from beneath—(from the retainer side), wind the pulley clockwise

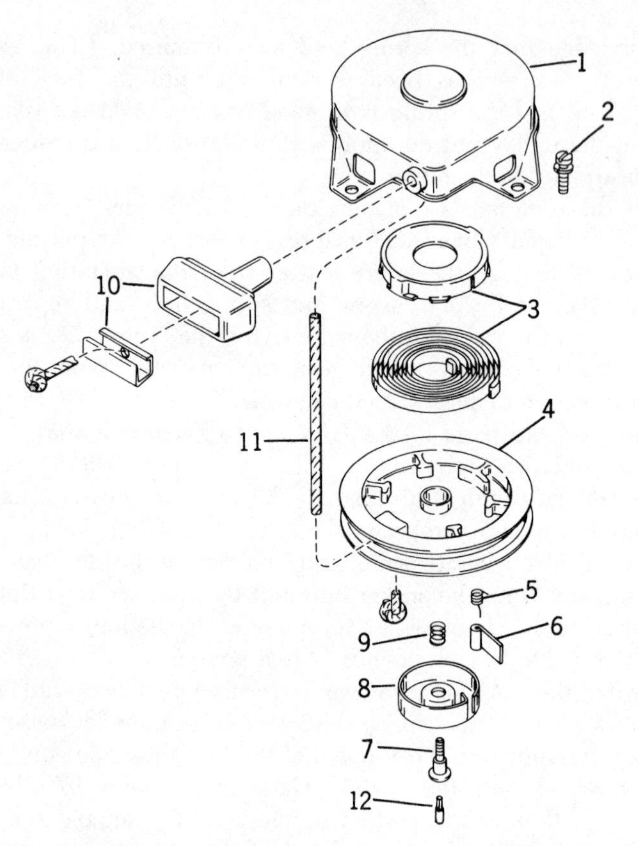

1—Starter Housing Assembly	*7—Retainer Screw*
2—Housing Screw, 1/4 x 3/4 (4 used)	*8—Retainer*
3—Spring and Keeper Assembly	*9—Brake Spring*
4—Pulley	*10—Starter Handle*
5—Dog Spring	*11—Starter Rope*
6—Starter Dog	*12—Centering Pin*

Dog-Type Recoil Starter Components

Illustration 5/10.

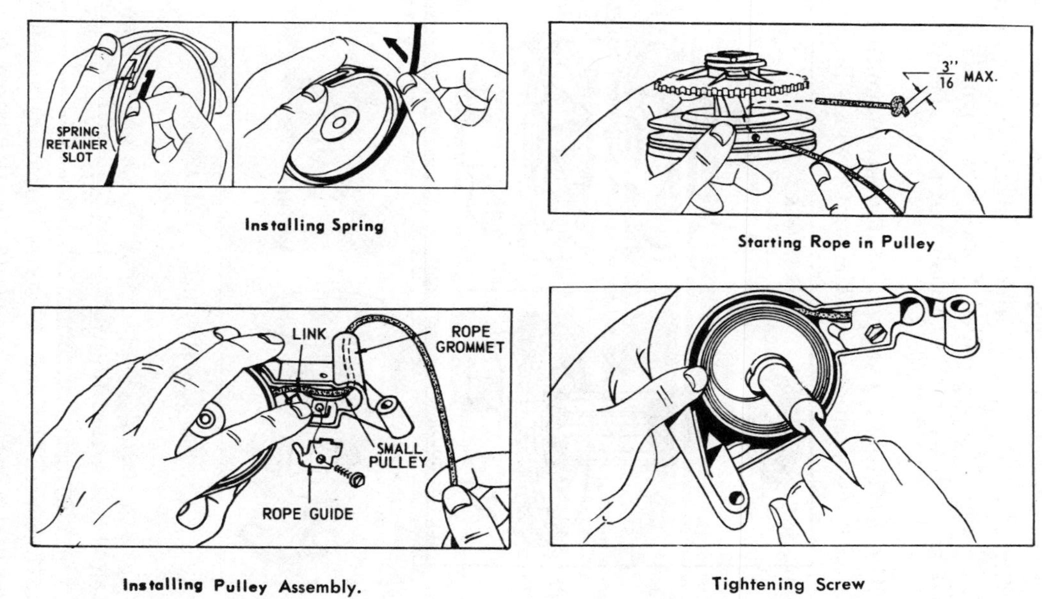

Installing Spring

Starting Rope in Pulley

Installing Pulley Assembly.

Tightening Screw

Illustration 5/11A-D. Reassembling recoil starter.

five or six revolutions, stopping with the pulley eye and housing eye in line so the rope can be easily slipped through them. Hold onto the pulley to retain the tension until the rope has been tied back into its handle. After the various parts are all cleaned and correctly reassembled, the starter dog should engage the flywheel when the rope is gently pulled. Conversely, when the flywheel is turned smoothly by hand, it should not push against the starter dog.

BRIGG & STRATTON rope pull starters are similar to Tecumseh in most essential aspects. The starter which Briggs calls its

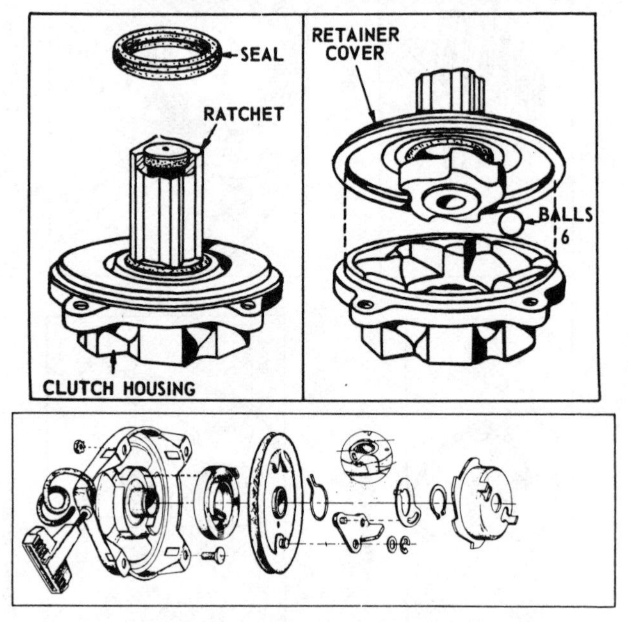

Illustration 5/12. Sealed Clutch Assembly.

"*vertical pull starter*" closely resembles the one used on the Tecumseh "LAV" series. Illustration numbers 5/11a through 5/11d depict peculiarities of the Briggs & Stratton system.

The Tecumseh "H" and "V" series starters are closely matched by the Briggs & Stratton *recoil starters.* Tecumseh recommends tightening the recoil spring 5 or 6 revolutions before a starter is

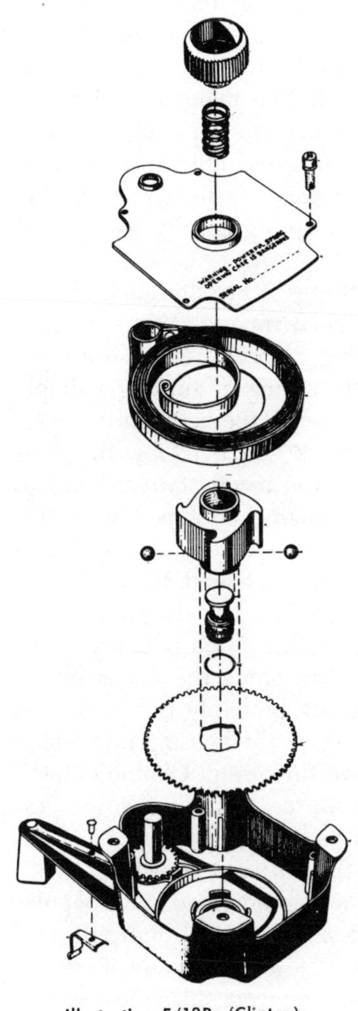

Illustration 5/13B. (Clinton).

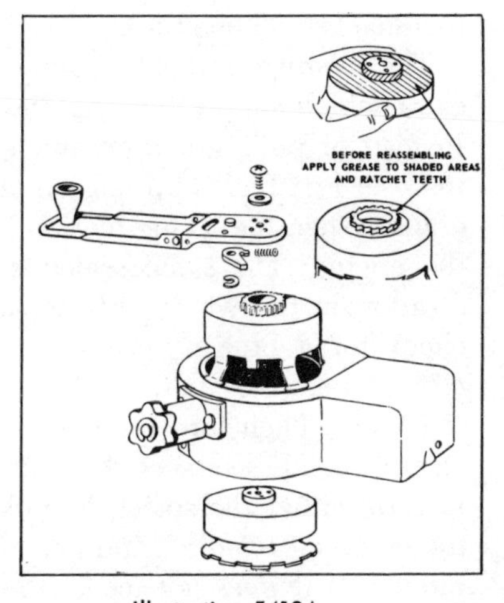

Illustration 5/13A.

Rewind Starter (Briggs & Stratton).

reinstalled whereas Briggs suggests 13-1/4 turns. To obtain so much tension, Briggs & Stratton tells repairmen to insert a piece of wood into the pulley hub and twist on the wood with a wrench for the necessary number of turns. Instead of dogs on the Briggs & Stratton starter, a sealed clutch engages the flywheel when the rope is pulled. Illustration number 5/12 shows the clutch sealed and opened. If repairs are required, however, the simplest remedy is replacement since the entire assembly cost is low. Clinton's version of the recoil starter is shown in drawing number 5/12c. Upon installation, the spring is fully wound and then backed off one turn.

The *windup starter* is a simple Briggs & Stratton and Clinton extra which simply reverses the normal rope starting sequence. Instead of pulling a rope and letting a spring retract the rope, the user first builds up tension on the spring by turning a simple crank. When the spring tension is released, the hefty spring spins the engine. The same sealed clutch is employed by Briggs & Stratton in the windup starter as in the recoil starter. You can check for a broken spring in your engine equipped with this type of starter by moving the start lever, wherever it may be, to "START." Then turn the cranking handle at least 10 revolutions clockwise. (Disconnect the spark plug first.) If the engine fails to turn, either the spring is broken or the clutch is faulty. Turn the cranking handle again and this time study the starter clutch ratchet; if it *does not* move, the spring probably is broken. Replace it, but *do not* ever attempt to open the housing into which the spring is coiled. It is a very powerful spring! Clinton's clutch also uses steel balls for its windup starter, but the spring is less formidable; its tension is lower and its housing is safer. But Clinton still recommends that you open the starter at arms length.

ELECTRIC STARTERS are becoming more and more popular these days, replacing the need to bend over a bit and pull on a piece of rope. Three basic designs have evolved, and these are found almost universally on all power mowers since most engine manufacturers generally buy the electric starter parts from factories who specialize in them. On smaller engines, a small 12 volt nickel-cadmium battery powers a small 12 volt electric motor for 3 or 4 second spurts until the gasoline power takes over.

Larger units with a 12 volt acid and lead storage battery use a larger 12 volt electric motor which can be run for half a minute or so at a time if necessary to coax gasoline power out of an engine. And some larger units employ a starter motor which plugs into standard 110 volt house current. You can turn over the 110 volt starter for a minute or so before the motor starts getting excessively hot from the effort.

There are very few parts in electric starting mechanisms which can be repaired by home handymen or even pros without at least a small set of electronic test equipment. If you feel up to such fiddling, you should buy a detailed repair manual for your particular mower engine. If you own an engine or a mower whose manufacturer or assembler keeps such service manuals in the east wing of Fort Knox, buy Briggs & Stratton "Repair Instructions IV." The electric starters comprehensively discussed there will more than likely be extremely similar to your own engine's starting mechanism. The repair manual for John Deere lawn and garden tractors will also prove to be valuable in servicing almost any starter as well as several Kohler and Tecumseh engines.

Illustration number 5/14 shows exploded views of typical large starting motors. Drawing number 5/15 illustrates the nickel-cadmium (ni-cad) system.

A BENDIX OR SIMILAR DEVICE is worked into each motor assembly as an integral part of the gear mechanisms. When the electric motor starts turning, the Bendix gear clicks into place on the flywheel gear teeth. Once the gasoline engine itself gets going, the starter gears have to be disengaged quickly. The Bendix, or its counterpart in other systems, is sensitive to the amount of pressure exerted upon it by both the starter motor and the flywheel. When the flywheel power matches that of the starter, the Bendix gear quickly snaps out of place.

The Bendix movement can be damaged after a "false start" if you attempt to turn on the starter motor again before the gasoline engine has come to a complete stop. Exploded drawings number 5/16a and 5/16b show how you might be able to check

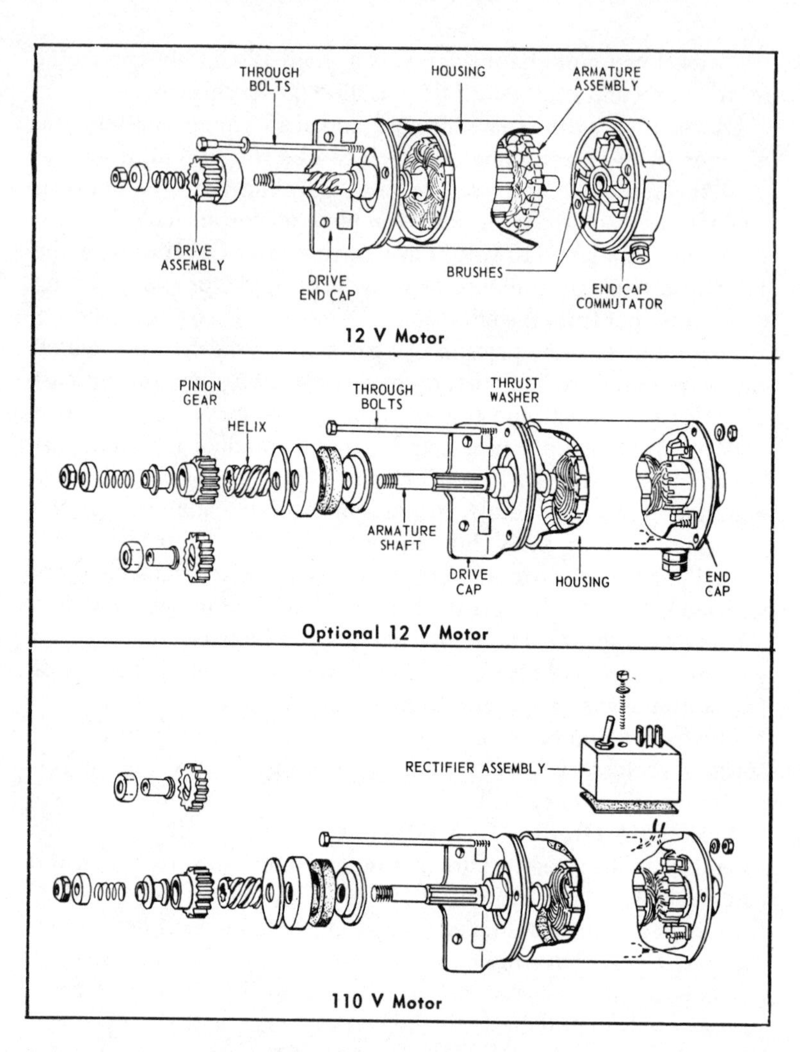

Starter Motors - - Exploded Views

Illustration 5/14.

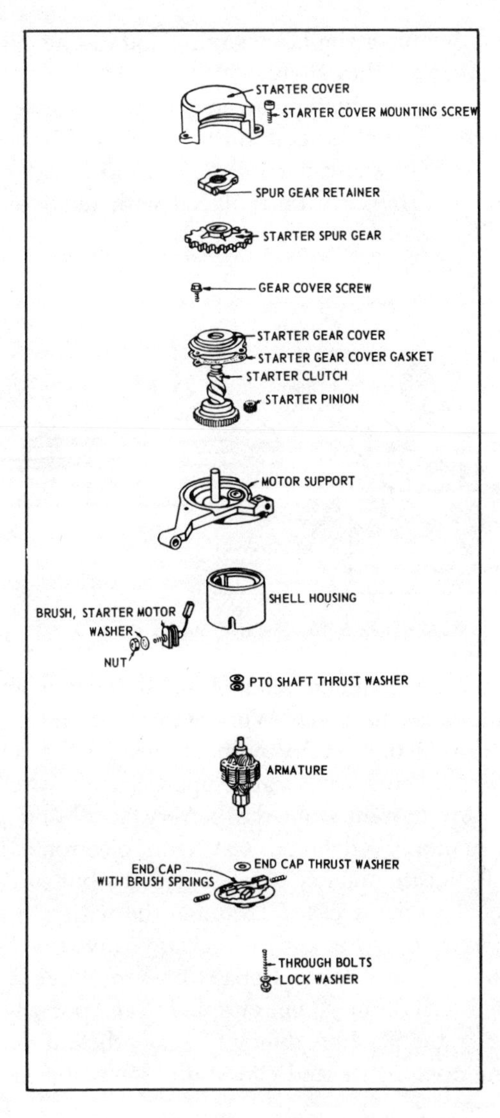

Illustration 5/15. Exploded view of typical starter motor for a Nickel-Cadmium battery system.

whether the Bendix or similar engaging and disengaging gears are functioning the way they should.

Aside from the Bendix type arrangements, *brushes* are the next most frequent ailment among starter motors. Illustrations number 5/14 and 5/15 show the location of the brushes in various starting motors. They can be replaced with relative ease and at extremely little cost.

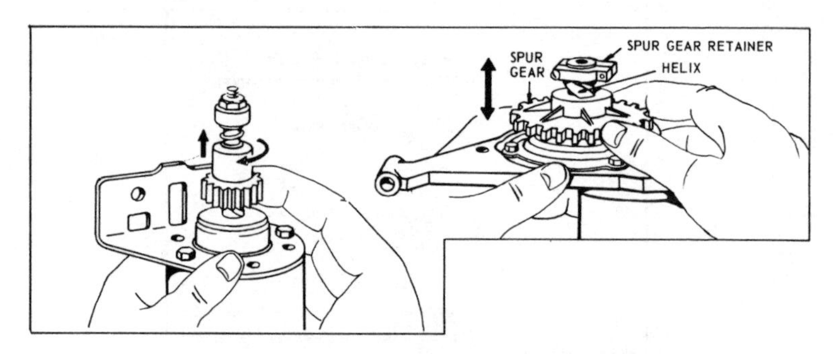

Illustration 5/16. Checking Starter Motor Drive.

The next most common ailment of starting motors is *worn wires* or *disconnected wires*. Worn wires can cause shorted circuits which drain the battery without turning the starter motor. And the disconnected wires cause open circuits which generally don't draw any current from the battery at all and do not spin the starter motor. A detailed test with electronic instruments can generally locate open or shorted circuits, but so can a simple physical check in many cases. Examine the wiring connection by connection, wire by wire, and correct any frayed or broken insulation with electrical tape. Repair broken wires or damaged connections by soldering them or splicing temporarily with extra wire to see if the starting difficulty really lies at that point. If you find one open or shorted condition, don't stop looking; there might be more.

Even the small ni-cad starting motor is significantly expensive, so it is well worth exploring a bit before rushing out for a new motor or a new mower. After all, that's what this book is all about.

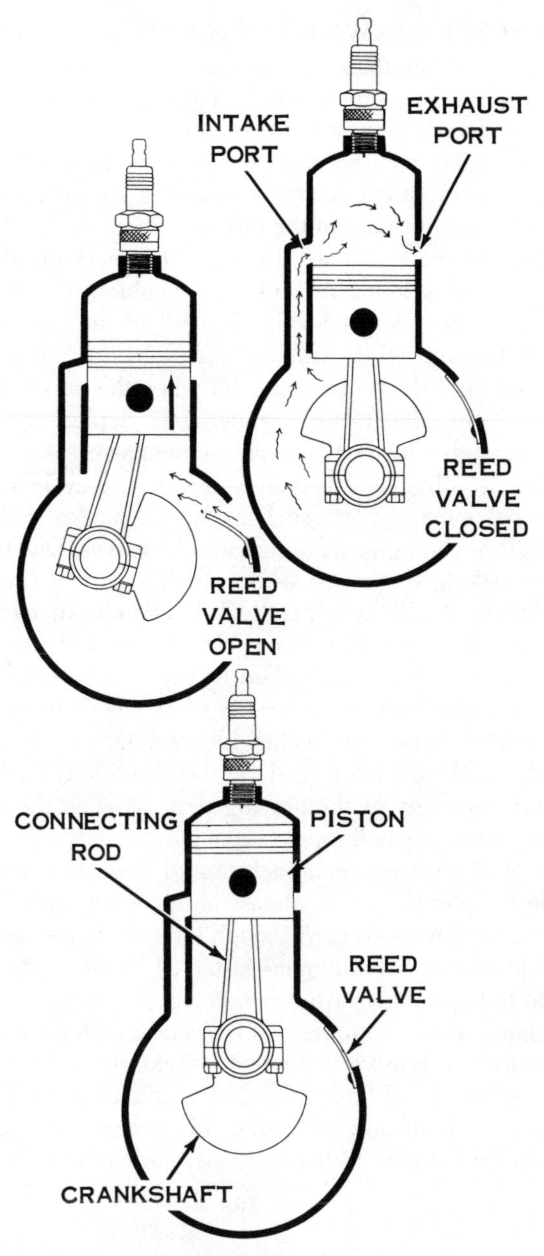

Illustration 5/17. Theory of the Two Cycle Engine. Fuel and air mixture enter via the reed valve. Exhaust is forced out the exhaust port at almost the same instant that new air and fuel is sucked in from the crank case.

TWO-CYCLE LAWN-BOY ENGINES operate on an entirely different principle than engines discussed thus far in this chapter and the preceding one. Instead of going through four complete cycles for every one power stroke, the two-cycle engine found on most Lawn-Boy mowers and some equipment sold by other lawn mower manufacturers, have one power stroke for every backward and forward movement of the piston.

On a two-cycle engine, there are no mechanically operated metal valves, as found in four-cycle engines. The two-cycle engine uses a flexible reed valve instead of an intake valve. The opening through which exhaust gasses are forced, is covered by the piston until the precise moment when the burnt gasses should be forced out. Illustration number 5/17 depicts the three basic stages through which a two-cycle engine passes.

The most obvious advantage of the two-cycle engine is its scarcity of moving parts such as no mechanical valves and no cam-shaft or timing gears to operate the valves. Due to the necessity for making the crankcase so much a part of the overall engine design, it also is impractical to include an inexpensive oil pump or oil slinger in a two-cycle engine. Lubrication of vital internal moving parts is handled by mixing oil with the gasoline. There is a tremendous advertising advantage in being able to claim that a two-cycle engine requires no oil change, but oil carried into the cylinder head via the gasoline is unable to reach efficiently as many of the moving parts as even the crude lubrication systems of small four-cycle engines.

One of the largest manufacturers of lawn mowers, Outboard Marine Corporation, uses almost nothing but two-cycle engines on its Lawn-Boy mowers. Although industry-wide sales and durability figures are virtually non-existent, Lawn-Boy mowers do not seem to fall apart any faster than competing brands. And another major lawn mower manufacturer recently added some two-cycle engines to its extensive line which previously included only four-cycle engines. So, although it is impossible to prove that the two-cycle engine is of any particular advantage over four-cycle engines except for advertising copy, neither can they be written off as inherently worse than four-cycle engines when used on rotary lawn mowers.

The two-cycle engines used on rotary lawn mowers are considerably smaller than the four-cycle engines in use. From Outboard Marine Corporation's own specifications, "A" Series engines barely hit 2 horsepower. Their smaller and medium sized "C" Series engines are rated about 2.5 h.p. to 3.0 h.p. The "D" Series and the larger "C" Series engines put out 3.5 h.p.

MAINTENANCE OF TWO-CYCLE ENGINES begins with always using the proper mixture of fuel and oil. Always use fresh gasoline and always use the recommended grade of oil, and in the proper proportions described in user manuals. Increasing the quantity of oil leads to poor carburetor performance, poor combustion and loss of power. Decreasing the amount of oil also will decrease the amount of lubrication which reaches the internal moving parts.

Cleaning carbon away from the exhaust ports is vital to good performance and long life in two-cycle engines. It should be done after every 50 hours of mower use or at the end of every mowing season. The spark plug must be removed or deactivated by tying the spark plug wire carefully out of the way. With a Lawn-Boy machine, the mower is tipped carefully onto its side and the muffler is removed, generally by unscrewing two bolts. The exhaust ports thus exposed will have an obvious accumulation of carbon around them. Pull gently on the starter rope until the piston covers the exhaust ports. Insert a 3/8-inch wooden dowel (which most hardware stores stock) into each port and vigorously loosen the carbon deposits. Lift the mower back to its upright position and pull the starter rope several times to blow the loose carbon out of the cylinder. Tilt the mower again and reinstall the cleaned muffler.

The ignition system in a Lawn-Boy mower is under the flywheel, but before probing into it, make certain that you have checked other potential sources of ignition difficulties outlined in Chapter 3. Ignition systems in a four-cycle and two-cycle system are virtually identical. On the small, simplest Lawn-Boy engines, called the "A" Series and "C" Series engines, it is rela-

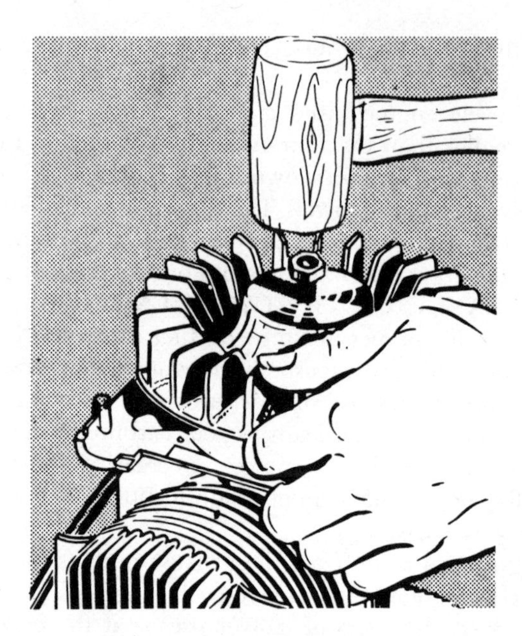

Illustration 5/18. Removing the flywheel on a Lawn-Boy engine. A soft hammer must be used.

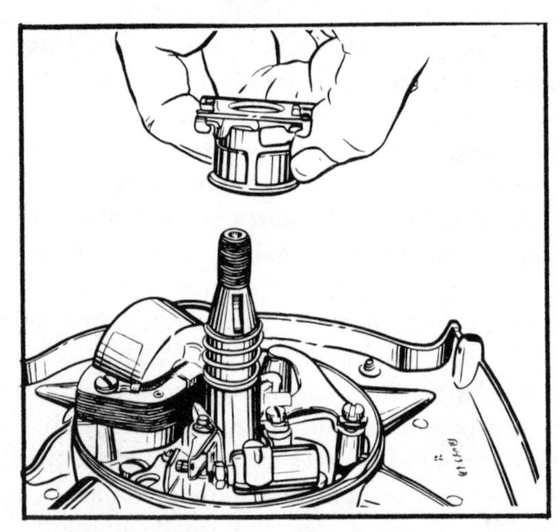

Illustration 5/19. Removing the governor from the crank shaft.

tively simple to expose the armature and breaker points. First remove the shroud and any other parts covering the flywheel assembly. Then remove the flywheel nut, washers, starter pulley and all related hardware. Next, screw the flywheel nut back into place until it is flush with the top of the crankshaft. With a *soft* hammer—hard rubber, wood or even a common hammer if you first protect the crankshaft and nut with a small block of wood—tap sharply on the nut while at the same time lifting up on the flywheel. It should slip off easily. Remove the flywheel key.

On the "C" Series engine, the governor is removed as a single assembly, as shown in drawing number 5/19, and then the governor spring. Larger Lawn-Boy engines, the "D" Series, have a more complicated governor to remove before you can service internal ignition parts. The "D" Series engine is not inherently more difficult to service, but it has more parts and is therefore more complicated. If you own one and wish to service it in detail, you can purchase the "Lawn-Boy Mechanic's Handbook" for a very reasonable price. The handbook also could be valuable to owners of smaller engines who wish to dig deeper into their engine.

Lawn-Boy says in its mechanic's handbook that "Coils and condensers seldom cause trouble—DO NOT replace a coil or condenser unless you are certain they are bad." The ignition points are replaced whenever they appear to be pitted or burned, or if supporting hardware or insulation looks worn. There is an oiling wick which lubricates the breaker cam; place a few drops of light oil on the wick whenever you expose the ignition system. Almost all breaker points are set at .020″ on Lawn-Boy ignitions.

On "A" Series engines, there is a timing adjustment screw. After the engine has been completely reassembled, let it run long enough to reach normal operating temperature. Then loosen the screw shown in drawing number 5/20 and twist the magneto plate just a few degrees left or right. When the plate is in the position where the engine runs most smoothly, tighten the screw securely.

"D" Series engines have a dust cover included over the magneto. And owners of "C" Series engines by Lawn-Boy can buy a separate dust cover for their magneto, a wise investment to make at the time of your first maintenance on the ignition.

Carburetors on two-cycle engines are generally very simple and attach directly onto the crankcase. When the moving piston

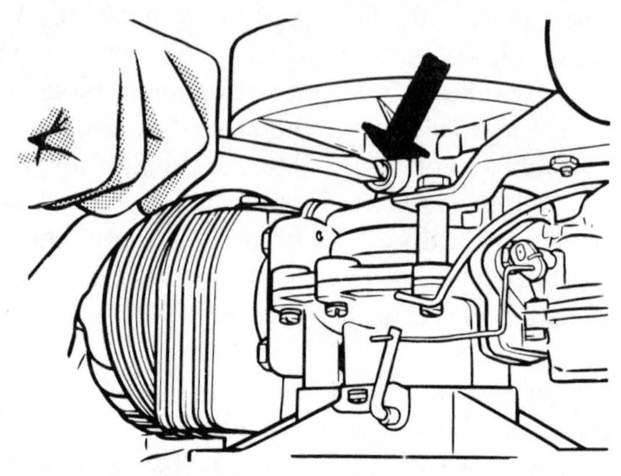

Illustration 5/20. Location of timing screw on "A" series Lawn-Boy engines. It is loosened and then the magneto plate is moved slightly left or right to obtain the smoothest performance. Finally, tighten the screw again.

creates a vacuum in the crankcase, the reed valves are pulled open and air is sucked in through the air filter. As the air passes through the carburetor, it sucks up a quantity of gasoline; the air and gasoline mixture passes through the reed valves into the crankcase. Drawing number 5/21 shows the usual layout of parts. Reed valves allow air to flow in one direction only, thus building up pressure inside the crankcase when the piston moves downward. It is that pressure which forces the fuel mixture into the upper cylinder.

Drawing number 5/22 gives an exploded view of a Lawn-Boy carburetor. To make sure that the nozzle is open, if trouble shooting tells you that carburetion appears to be a problem, twist the nozzle valve fully but gently shut. Then open it two full turns and attempt to start the engine. If it will not start, gradually open the valve several more turns. If the engine does not start after the nozzle is opened to five complete turns, the trouble lies elsewhere, perhaps in a gummed up orifice inside the carburetor. Solvents can be used to clean out this carburetor, but

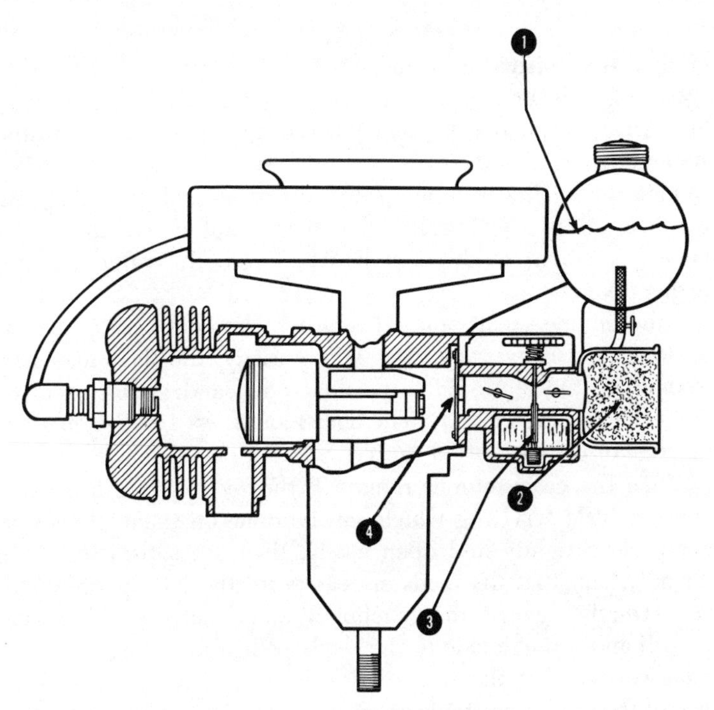

Illustration 5/21. Lawn-Boy carburetion system. (1) Fuel Tank which feeds gas by gravity. (2) Air cleaner. (3) Nozzle through which the flow of air sucks gasoline from the reservoir regulated by the float which surrounds the nozzle. (4) Reed valves which allow the air-gasoline mixture into the crankcase.

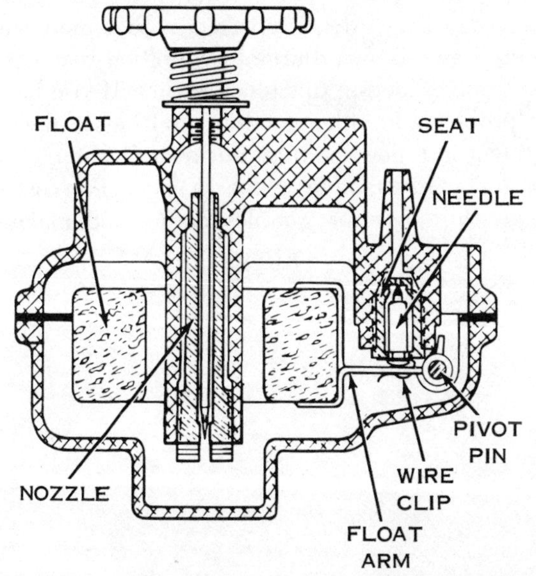

Illustration 5/22. Cross section of Lawn-Boy carburetor.

the float is varnished cork and must be kept away from solvents; any strong solvent which gets on the float probably will dissolve the varnish and make the float porous, thus changing its floating characteristics.

Once the engine is running and up to normal operating temperature, adjust the nozzle valve to a point where the engine seems just about ready to stall. Then turn the valve 1/2 turn further open.

If the engine seems starved for fuel, the float setting may be too low. And conversely, if it floods easily, the float may have become set too high. With the float bowl and gasket removed, invert the carburetor body. The float should rest 15/32-inch above the edge of the carburetor body.

When the carburetor is removed, the reeds are also exposed. There is little servicing which can be done on them. If they are clean, close tightly and open easily, they are satisfactory. It is not necessary that the reeds appear perfectly flat, but they must close snugly against the machined metal surface. There is a smooth and a rough side to the reeds. When installing a new set, make certain that the smooth side is toward the machined surface of the carburetor body.

Newer Lawn-Boy mowers feature modular construction. Both carburetor and ignition systems are one-piece. Neither have any adjustments. Either they work or they don't work, and in the latter case, they have to be replaced with a new module. In servicing the Lawn-Boy carburetor or ignition modules, you can follow the troubleshooting tips given earlier. If you have tracked down the problems to either module, you face a fairly substantial investment in a new module. But overall, the cost should be much less for a new module alone than for the cost of a new module *plus* the labor for troubleshooting and making the replacement.

CHAPTER 6

Rotary Lawn Mower Repairs

BASICALLY there is only one moving part on simple rotary lawn mowers—the engine itself. Such simplicity could lead you to conclude incorrectly that rotary lawn mowers last forever. Alas, the engine has to operate at consistently high speeds and under adverse conditions such as vibration and sudden shocks from hitting rocks, bottles and clumps of dirt. Key engine parts like bearings and crankshafts wear out relatively quickly under these circumstances. The parts which do wear out on an engine mounted atop a rotary mower are often the very ones which you may find the most difficult to replace or to repair without specialized tools and well formed mechanical instincts.

Your best defense in contending with a rotary lawn mower is to minimize the likelihood of ever needing major repairs. The blade must be kept sharp and well balanced. You must scour a lawn before cutting it to remove obstructions like rocks, bottles, branches and even clumps of dried and matted grass from past mowings.

Mounting bolts for the engine must be checked frequently to insure that they haven't rattled loose. The blade mounting bolts and adapters have to be checked before every mowing, both for safety and to avoid excessive engine vibration. Self-propelling mechanisms should be kept well lubricated and well adjusted so that starting and stopping proceeds smoothly and will not add additional jerks and bounces to the crankshaft and bearings.

Even the condition of the deck itself is important in rotary mowers. A cracked or distorted housing over the blade can be dangerous if there is any chance that the blade could strike any part of the damaged deck. Since the deck separates the engine from the blade almost like a bearing, its design and maintenance can itself lead to additional harmful vibrations.

SHARPENING A ROTARY LAWN MOWER BLADE IS RELATIVELY SIMPLE. A hand file can do a nice job but since this is a power-hungry age, you can switch to one of the many power sharpeners available. Simplest of all is a low-cost gadget which slips into a 1/4-inch electric drill. The proper kind of drill attachment fits over the blade edge; there is a sharpening wheel for the top and a support wheel for the bottom. One or two slow passes with a drill-powered sharpener often is enough to hone a blade which isn't badly knicked.

Bench grinders found in some home workshops these days can do a fine job of reshaping a bent, nicked and dulled edge. If you make more than two or three passes across the blade, use a generous application of cutting oil. Otherwise the blade edge could become so hot that the heat will begin to dull the blade before the knife edge ever is finished.

Whether you use a drill or a bench grinder, whenever a grinding wheel is used, *protect your eyes* with goggles. Those companies which cheerfully sell you a two dollar grinding wheel mechanism for a drill should be required by a sense of morality if not by law to include a set of plastic goggles. Grinding wheels themselves occasionally will shatter and fly apart, especially if dropped or hammered. And a severe dent in a blade might not be ground away gradually as you run the blade through the grinding apparatus—the knicked portion might simply break off and fly at you.

The dangers of sharpening blades should not be *over-exaggerated.* It can be done with reasonable safety if you don't mistreat the grinding wheels and if you wear a cheap pair of plastic goggles over your eyes or over your glasses.

Before attempting to sharpen a blade, inspect it for cracks or breaks. Such a blade must be replaced because of the danger it represents on your mower. A simple test can help you pick out damaged blades. With the blade suspended in air, tap it with a bolt or similar metal object. It should almost tinkle like a bell. If you hear a dull thud, suspect a crack or other structural fault.

A rotary lawn mower blade should not be sharpened like a knife. A blade which comes to a sharp, pointed edge is very prone to damage by even small foreign objects. Not only is it sensitive to being nicked, the thin edge can be rolled over; you'll soon have a blade with a greatly reduced cutting surface. Approximately 1/64″ of the cutting edge should be flat. Although the flattened cutting surface is not as sharp in the beginning, it

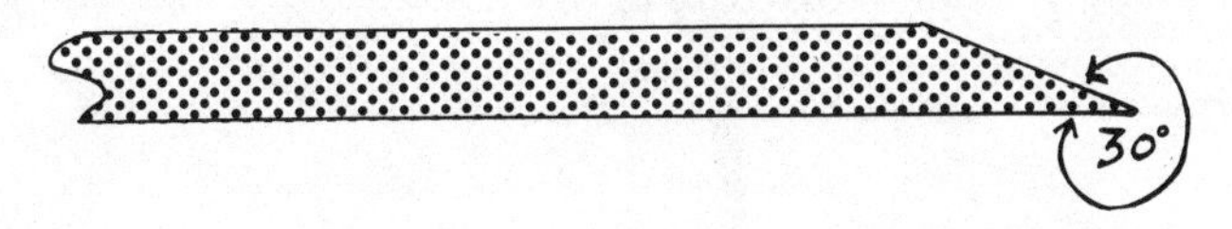

Illustration 6/1. Incorrect (top) and correct (bottom) cutting edge for a Rotary Mower Blade.

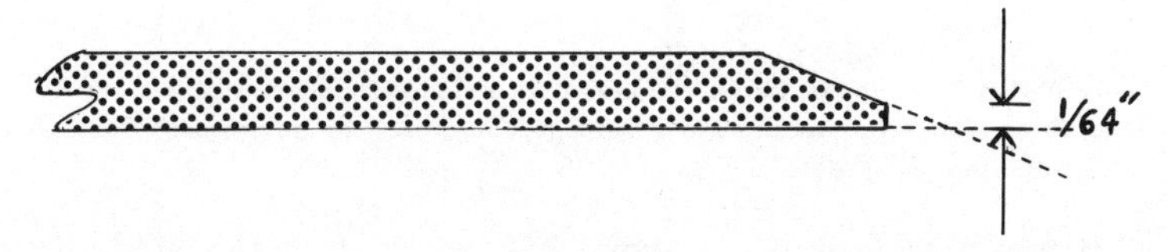

retains its sharpness much longer. By the time your cutting season is only half over, the "dull" edge should be considerably sharper than the blade which began with a very pointed edge. For the technically minded, the slope of the cutting edge is kept close to 30°, as shown in drawing number 6/1.

A BLADE MUST BE FINELY BALANCED in addition to being sharp. If one end of the blade weighs even slightly more than the other, it can generate tremendous vibration by the time the engine reaches its full speed of maybe 3000 r.p.m. One testing laboratory reports that objectionable vibration is created when technicians deliberately unbalance a 20″ blade by a fraction of an ounce.

A cone shaped object of the proper size, acting as a fulcrum, is the most satisfactory balancing device available for non-professional use. Some lawn mower supply shops sell an inexpensive fulcrum which will handle the wide range of center hole sizes found on the market. The blade center holes range from 3/8″ to over 1″ in diameter. For smaller sizes, the caps of some marking pens or thick pencils can be used for testing blade balance. The necks of soda pop bottles or other bottled refreshments can balance blades with very wide center holes.

Illustration 6/2. A Rotary Blade is checked for balance by balancing it at it's center hole. If the blade tips noticeably in either direction, it must be balanced by filing away metal from the heavy side.

Before getting into the actual balancing, clean the blade since any significant accumulation of dirt will have an effect on the blade balance. Once the horizontal blade is resting on the vertical balancing device, look carefully to see if the blade tips noticeably in either direction. If so, you must grind or file enough metal off the heavy side of the blade. Since the cutting edge already has been perfected, remove enough metal from the back side of the blade's heavy end to balance it exactly.

After the blade is back on the mower, perform one more check which can cut down on engine wear. (*Make sure the spark plug wire is disconnected and tied down a safe distance from the spark plug.*) Hold a ruler against the deck and place the end of the ruler against one tip of the blade. Now, being careful not to move the ruler, rotate the blade and measure the distance to the second tip. There should not be more than 1/16″ difference. Check to see that you have installed the blade properly, that the various shear pins and bolts are all in their assigned places and that the bolts are uniformly tightened. Perform the measurement sequence again, and if the blade still shows more than a 1/16″ bend or bow to it, you may have to discard that blade. Drawing number 6/3 graphically demonstrates this test.

All whirling blades generate intense amounts of centripetal force. Blades which are perfectly straight or uniformly shaped

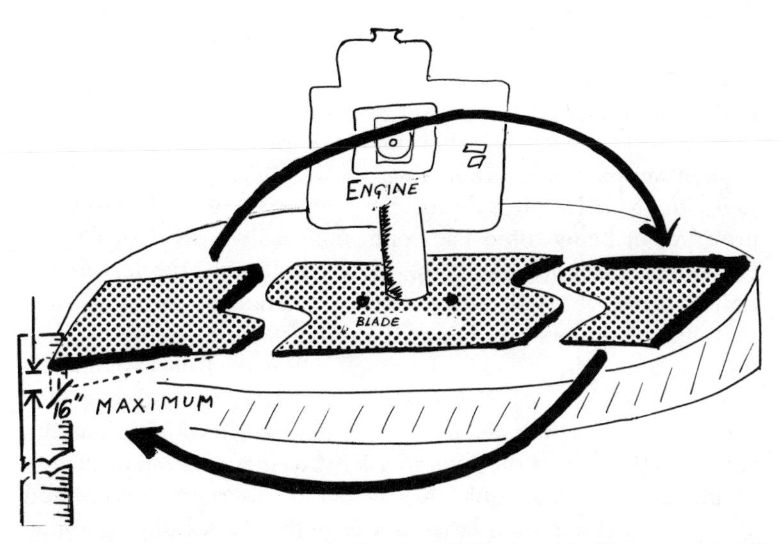

Illustration 6/3. Checking that a Rotary Mower Blade is straight. More than a 1/16″ bend can generate vibrations and other undesirable forces which can be potentially harmful to the user or to the engine.

on both ends, generate the centripetal force evenly on all parts of the crankshaft and its bearings. Let the blade bend or twist more than 1/16″ away from its true straightness, however, and the centripetal force tries to bend the crankshaft into an orbit with the bent blade instead of the orbit which the engine manufacturer gave it. The result, over a period of time, can be a bent crankshaft or a set of worn bearings. It is quite significant that a warning about this particular problem was found in user literature supplied *by only one lawn mower assembler,* Deere and Company.

THE DECK OR BODY OF THE ROTARY LAWN MOWER is generally made of aluminum or magnesium alloys, ostensibly to make them as light as possible. This appears to be logical since human-power is needed to push a non-self-propelled machine. Those lawn mower assemblers who do offer steel decks, however, generally do so on the top-priced models in their line. A sensible person must wonder whether pounds or profits dictate the use of lightweight decks. There is only a two pound difference between a magnesium deck and a steel deck.

The most common explanation for damaged decks, aside from faulty stamping or casting by the assembler, is that the user attempts to turn some very unorthodox corners. *One wheel* is not enough support for a thin lawn mower deck, and neither are *two side wheels.* Most bases are reinforced to withstand the pressures of being tilted back onto the rear wheels. However, if you use the handles to tilt the mower onto only the left or the right wheels, you will be adding more strain to the base than most manufacturers ever hoped they could withstand.

A crack or split of only a few inches in a steel deck should be welded. (You can tell if it is steel, of course, by testing with a magnet.) If you first remove the engine via its four mounting bolts and then carry the base to a local welder, the repair charge should be very reasonable. No welder in his right mind would try his arc or torch on a lawn mower with a tank full of gasoline or gasoline fumes only inches away. Although aluminum *can* be welded, the job might be beyond the equipment and experience found in many local welding shops.

A new deck can be expensive, but the fiberglas kits for repairing boats or dented auto fenders can also fix your aluminum or magnesium alloy mower housing. First you must scrape away all of that fancy paint. Paint remover can help if you're doing the job by hand. Otherwise a wire brush attachment for a power drill will accomplish the job. Get down to the bare metal for two inches on all sides of the split.

Before mixing the resin which comes with the fiberglas kit, cut a series of four patches from the fiberglas fabric itself. Cut the first patch one inch longer than the crack or split and one inch wide. In the case of a three inch split, for instance, patch number one would be one inch wide and four inches long. For successive patches, add one inch onto both the width and length. The fourth patch in our example of a three inch split would be four inches wide by seven inches long. If structural parts interfere with following these dimensions precisely, cut out a portion of each patch to correspond roughly with the dimensions of that interfering part.

Make sure that both sides of the split are level. Then swab on a generous portion of the patching resins to the metal. Apply the smallest patch; smooth it into place and dab on more resin as needed. Apply the second patch, smooth it into place well and again add whatever resin is needed to saturate the working surface. Patch number three and then number four are applied in the same manner. Care must be taken that each successive patch makes very good contact both with the patch beneath it and a sizeable strip of bare metal.

Since the resin-soaked fiberglas is very pliable, this repair technique need not be limited to flat deck surfaces. Once the resin has cured for the proper length of time specified by your patching kit, the cracked deck should be durable enough to provide about as much service as the original deck. The resins will protect the once exposed metal areas from rust but you can repaint the deck and patch so the neighbors will never know.

THE LIFE SPAN OF A TYPICAL ROTARY LAWN MOWER IS OFTEN ONLY TWO YEARS. Practically no representative of any lawn mower company has denied that. Principally it is the engine which wears out, or more appropriately, gets worn out by having to spin faster than it was intended to go and under harsher conditions than it was designed to withstand. If your engine gets beyond reasonable repairs, leaving you with a perfectly sound set of handles, wheels, deck and perhaps a self-propelling mechanism, replacing the engine might very well be more economical than buying an entirely new lawn mower.

Briggs & Stratton, for instance, offers new 3.5 h.p. engines from regional warehouses. They cost about half the price of a new rotary mower. From its Milwaukee factory, Briggs & Stratton can supply you with an engine to replace the special-order ones sold to Sensation, Jacobsen, Toro, Simplicity and on down the line. Engines for tractor style mowers also are available, usually costing less than one fourth the price of a new riding mower.

Assuming you supply the correct model and type number of your old engine and the proper replacement is furnished by Briggs or some other engine manufacturer, installing the new engine will be mainly a matter of unbolting the old one and bolting on the new.

SELF-PROPELLED ROTARY LAWN MOWERS COME IN TWO BASIC TYPES. One uses the ordinary vertical crankshaft engine to which the lawn mower assembler adds a pulley which connects to another pulley which indirectly or directly drives the wheels on your lawn mower. To accomplish this, the assembler has to use an engine with a longer exposed crankshaft than otherwise would be necessary or desired. The longer crankshaft exposure can bend easily. A pulley and belt attached to the exposed part of the crankshaft add just one more complicating element which can generate vibration and shorten the life of expensive engine bearings and of the crankshaft itself.

The second basic type of self-propelled mower simply couples the wheels via a clutch to the power take-off shaft of an engine and leaves the main crankshaft untouched except for the delicate mower blade itself. For several dollars more, lawn mower assemblers can buy an engine which has a power take-off built inside of the crankshaft housing. The power take-off shaft comes out with its direction of rotation already geared to the right position for powering the wheels. This is accomplished with a precision which most external devices cannot match.

Some of the self-propelled rotary mowers are self-propelled by a belt, some by a chain. Some use the handle to activate or deactivate the clutch and others employ a separate lever. If the handle acts as the clutch lever, which is the American National Standards Institute's approved method, it is difficult to adjust the height of the handle to accommodate people of varying body height. On the other hand, should a separate lever serve to

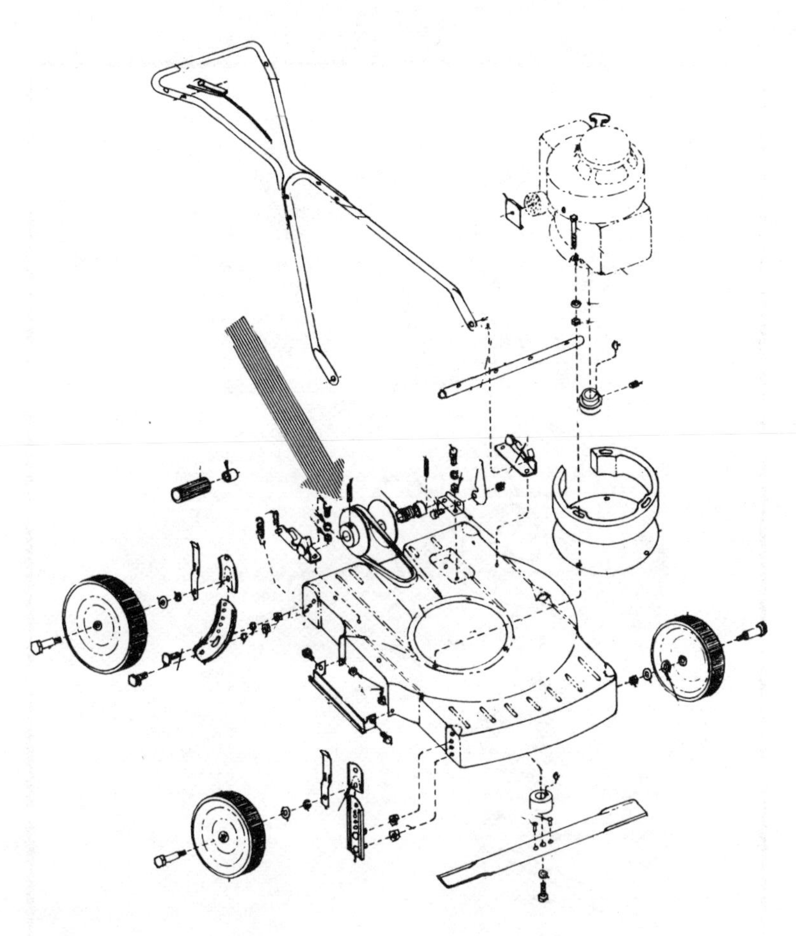

Illustration 6/4. If a pully provides self-propulsion to a mower, the belt must be properly installed to prevent the mower from running backward.

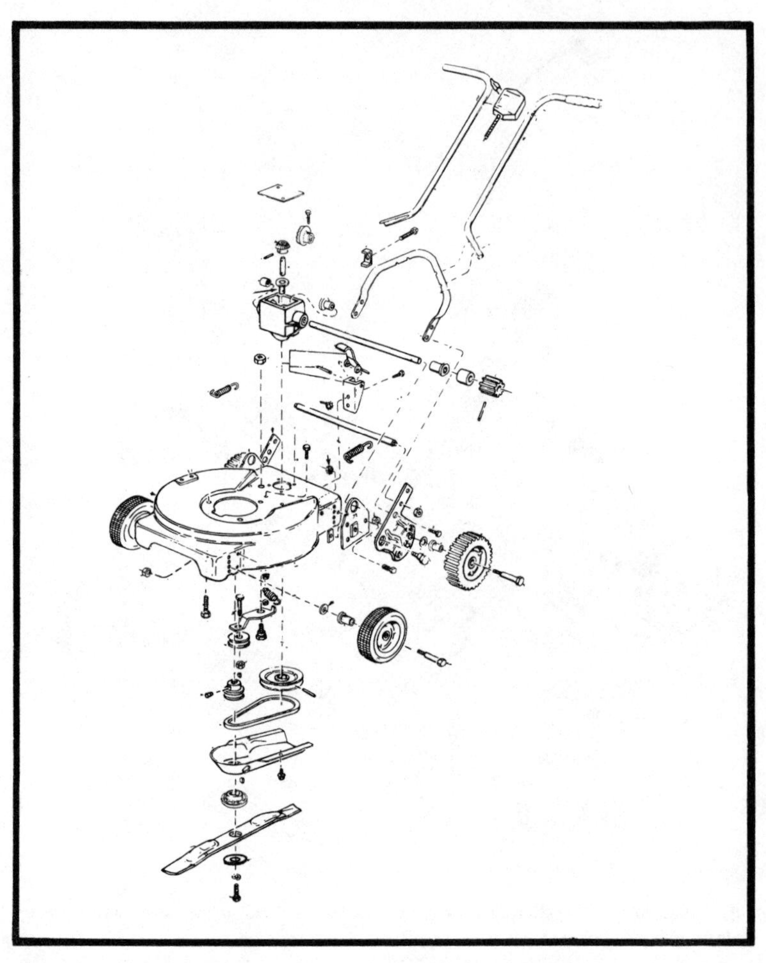

Illustration 6/5. The handle engages or disengages the self-propelling mechanism in this typical mower and likewise for the mower in Illustration no. 6/4.

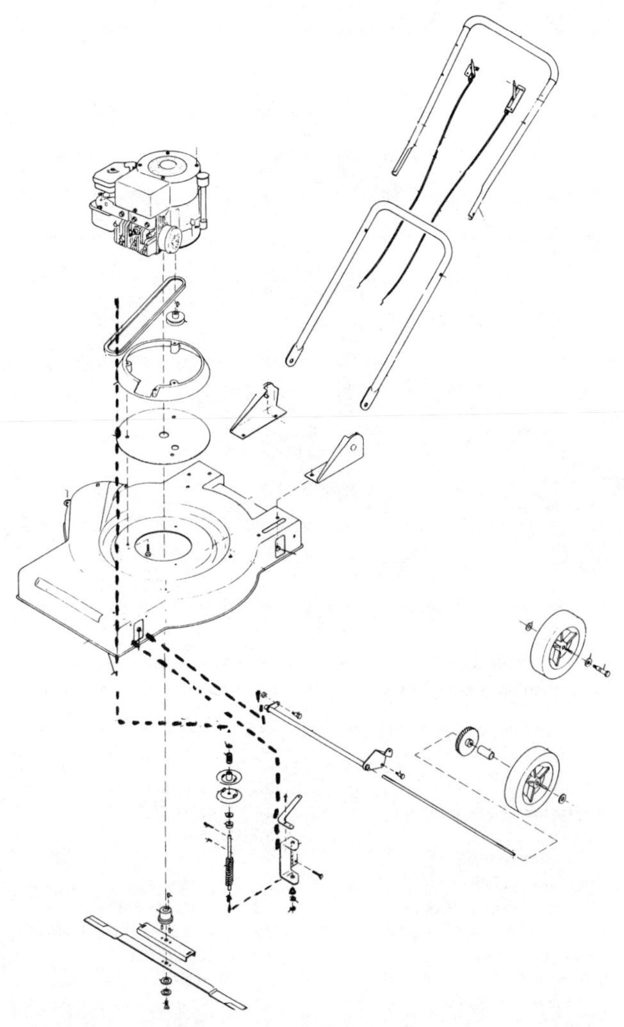

Illustration 6/6. A pulley-driven self-propelled Mower. The self-propelling mechanism is driven by a power-take-off shaft on the engine.

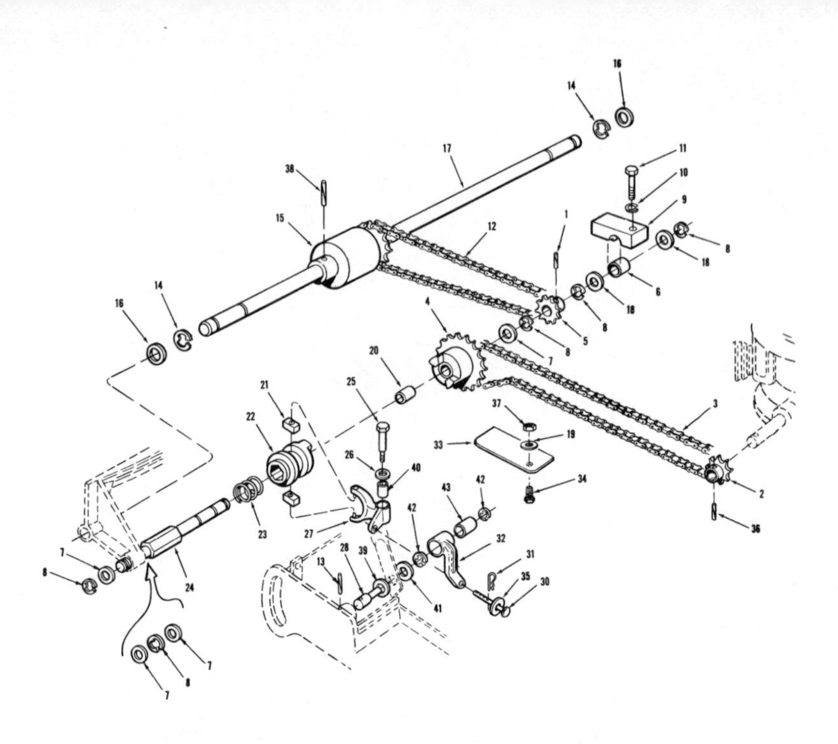

Illustration 6/7 Chain-Driven, Self-Propelling Mechanism
The mechanism is powered by a horizontal power-take-off shaft on the engine.

Ref. No.	*Name of Part*	*Ref. No.*	*Name of Part*
1	Rollpin	23	Clutch Spring
2	Sprocket (13 Tooth)	24	Jackshaft
3	Chain	25	Bolt—Shoulder
4	Sprocket Assembly	26	Washer (5/16 x ⅝ x 1/16)
5	Sprocket (13 Tooth)	27	Yoke
6	Sleeve Bearing (Bronze)	28	Handle Pin
7	Washer (7/16 x 1 x 1/32)	29	Washer (½ x 1 x ⅛)
8	Retaining Ring	30	Clevis Pin
9	Retainer	31	Hairpin
10	Washer (¼ Split)	32	Clutch Bracket Assembly
11	Bolt (¼-20 x 1⅛)	33	Plate
12	Chain	34	Screw (10-32 x ½ Phillips)
13	Rollpin (3/16 x 1)	35	Washer—Flat (11/32 x ⅞ x .075)
14	Retaining Ring	36	Rollpin
15	Sprocket Flex Coupling	37	Nut (#10-32 Hex.)
16	Washer (19/32 x 1¼ x 1/16)	38	Rollpin
17	Rear Drive Shaft	39	Washer (½ x 15/16 x 1/16)
18	Washer—Special	40	Bearing
19	Washer (#10 Split)	41	Washer (⅜ x 1 x .035)
20	Sleeve—Bushing (Bronze)	42	Retaining Ring
21	Shoe	43	Bearing
22	Clutch—Dog		

activate the clutch so your handle height adjusts easily, it is possible for the mower to keep moving with no one at the controls.

There are front wheel drives and rear wheel drives. Some of the drive wheels are rubber. And some are just overgrown gears. You can imagine what *they* might do to a new lawn and a somewhat soggy or delicate turf.

At least one self-propelled mower requires a belt which twists at a right angle between one pulley mounted on the vertical crankshaft and another pulley mounted on the horizontal axle. If you should perchance own one of those twisted-belt vehicles, make certain that you do twist the belt in the proper way when it needs replacing. A belt can be twisted 90° to the right, which would move the wheels in one direction, or twisted 90° to the left, moving the wheels in the opposite direction. The wrong kind of twist would send those potentially deadly blades spinning backwards right at you!

Drawings on near by pages show some representative self-propelling mechanisms. Number 6/4 shows the potentially damaging twisted pulley mechanism. For just a very few dollars worth of additional parts, the lawn mower assembler can take an otherwise sound rotary mower and turn it into a more glamorous but more dangerous self-propelled machine. The mower shown in drawing number 6/5 also taps the power of the main crankshaft with a pulley. This pulley goes to a small gearbox which changes vertical rotation into horizontal. In both drawings, number 6/4 and 6/5, when the lawn mower's handle is raised, the drive gear meshes with the geared rear wheel.

A vertical power take-off crankshaft on the engine shown in drawing number 6/6 drives the front wheels first via a pulley and then through a worm gear which slows down the speed and changes vertical rotation into horizontal. Instead of a pulley on the drive wheel end of the self-propelling mechanism, a set of two sheaves forms a "pulley." The sheaves can be loosened or tightened by a lever on the mower's handle, to stop or start the wheels.

A horizontal power take-off shaft is used in the self-propelled mechanism of drawing number 6/7. Notched "dogs" attached to the shaft which has two separate sprockets, are engaged by a claw-shaped arm. The arm has to pry the dogs apart with a left-to-right action while the handle moves through a front-to-back action. The entire operation is just as tough to accomplish mechanically as it is to describe in print!

Some of the more sophisticated self-propelling systems have what is known as an idler arm to halt the blades within 3 to 10 seconds after a user disengages the clutch. Such a feature adds greatly to the safety of a rotary mower. Drawings number 6/8 and 6/9 show typical adjustment settings for an idler.

Need it be said after the preceding several pages, that there is almost no standardization in the design of self-propelling rotary lawn mowers. It would be futile to offer here much more than a wish for good luck and a check list of basic areas to explore when a self-propelled rotary mower needs attention.

• *Engine connection:* Does the pulley or gear which links the self-propelled mechanism to the engine move when the starter is slowly twisted by hand? (*Disconnect the spark plug before attempting this test!*) If not, the pulley or gear may not be soundly connected. Or, in the case of an engine featuring a power take-off shaft, the gears inside the crankcase may have broken.

• *Link between engine and speed reducer:* In many instances, the high speed of the engine is reduced to the low speed needed for propelling the lawn mower by the pulleys or chains connecting the engine to the wheels. A small pulley on the engine and a large pulley on the drive wheel reduce the number of revolutions per minute. If the self-propelling mechanism is not working properly, however, the belt or belts may be slipping, or the various sprockets in a chain linkage may be loose. It is also possible that the chain has jumped off the sprocket and must be reinstalled, generally with a screwdriver.

• *Speed reducer:* A few machines use a gear box to reduce the engine's speed to a manageable number of r.p.m.'s for driving the wheels. If all of the connections leading into the gear box

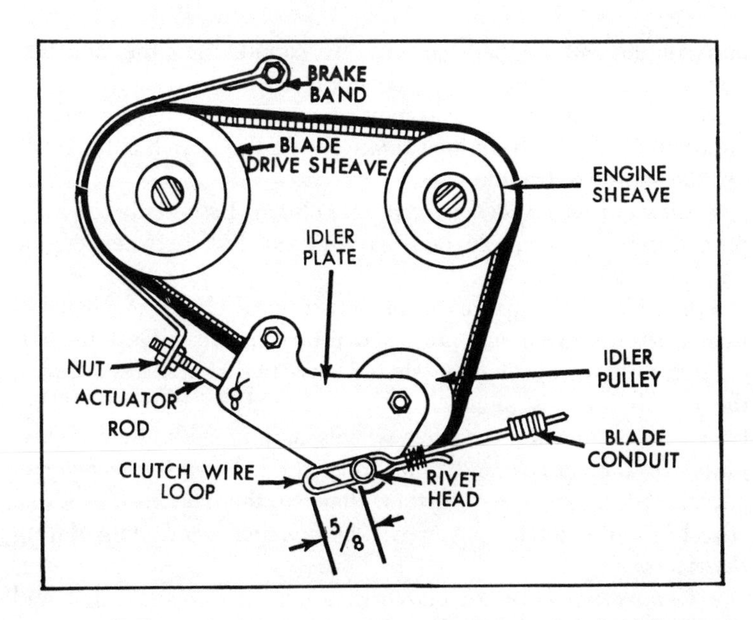

Illustration 6/8 and 6/9. Typical idler Arm Assemblies which halt blade rotation shortly after the power is turned off. Otherwise blades can coast for 15 seconds or more.

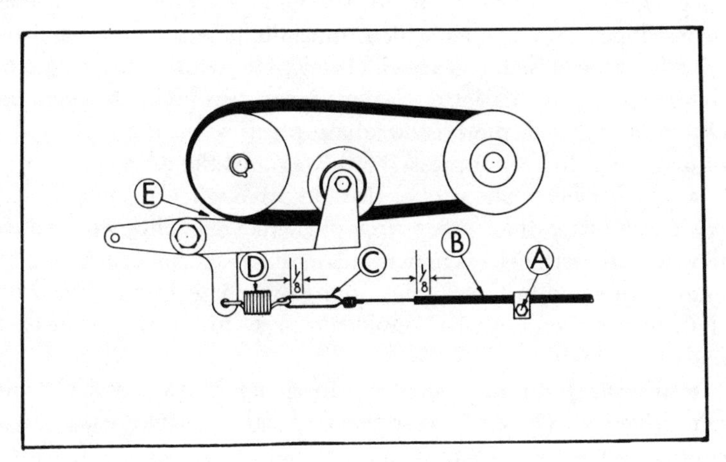

and coming out of the gear box are sound, then the difficulty may lie inside the gear box. Generally there are very few individual gears inside the gear box of a self-propelled mower. A handyman can carefully investigate inside the gear box to search out the source of trouble.

• *Link between speed reducer and clutch:* Essentially the same procedures should apply here as with the link between engine and speed reducer.

• *Clutch:* Earlier pages dealt with typical clutch mechanisms which are found on self propelled rotary mowers. Despite their diverse designs, all of them should pass two simple tests: when the clutch lever is *engaged,* power should be transmitted between the engine and the drive wheels, assuming all other moving parts work properly; and when the clutch lever is *disengaged,* power should move no further than to the clutch—the drive wheels should not be powered. (*Disconnect spark plug during these tests.*)

• *Connection between clutch and wheels:* The sprocket and chain or pulley and belt links require essentially the same examination as detailed above in the "link between engine and speed reducer" section. However, some self-propelling mechanisms depend on a roller which transmits power to rubber drive wheels or on a small gear which engages with a large, geared drive wheel. In this latter design, make certain that when the clutch is engaged, there is adequate pressure on the drive roller or drive gear to keep it pressed firmly against the drive wheel.

• *Drive wheel:* The rubber tires on drive wheels generally will wear out faster than other tires on a mower. This will reduce power and perhaps even neutralize the drive mechanism altogether. In some instances, the drive power is exerted on a drive shaft which then must be coupled securely to the wheels at each end of the shaft.

• *Wheels for rotary mowers:* In general, drive wheels and non-drive wheels can be replaced with relative ease. Most simply bolt onto a frame or directly onto the deck itself. By removing the bolt, you can take off the old wheel and slip a new one into its place.

CHAPTER 7

Reel Lawn Mower Repairs

COMPARING a reel type lawn mower to a rotary lawn mower is comparable to standing a Rolls Royce alongside an army tank. The tank and the rotary lawn mower operate on sheer horsepower; the gasoline engine or electric motor spins as fast as it can all the time and just mows down everything in its path by brute force.

A reel mower is a refined piece of machinery. The engine doesn't have to operate at full power and consequently generates less vibration, less noise and less danger of excess heat inside the engine itself. The reel mower has half a dozen or so blades clipping grass and a cutter bar which holds every blade of grass in place while the blade slices it at a uniform height.

A 21-inch self-propelled reel mower can do nicely with a 2 h.p. engine running well below top speed. The same size rotary mower requires at least a 3.5 h.p. engine going practically full throttle.

Like most fine pieces of machinery, the reel mower has more moving parts than its rotary mower counterpart. However, servicing a reel type mower becomes relatively simple once the fundamentals are absorbed. The engine is subjected to less vibration, heat and constant high speed operation so the piston rings, valves and bearings—the hard to service parts—need less attention. The governor no longer is a critical item on the reel mower because a potentially lethal blade is not spinning just inches away from the operator's body.

BASICALLY, THE REEL MOWER ENGINE is attached to a small pulley which drives a larger pulley via a v-belt. This larger pulley is in turn coupled to a sprocket, similar to the rear chain sprocket of a bicycle. This small sprocket drives a chain which spins a larger sprocket which generally is fastened to the lawn mower reel. As the reel spins, it turns a small gear called a pinion which drives the wheels via a larger gear built right inside the wheel itself.

Why doesn't the engine drive the reel more directly, say with a very small pulley or gear on the engine and a very large pulley or gear on the reel? First, by going through two successive speed reductions a smoother operation is possible. And secondly, a pulley allows lawn mower assemblers to design a very simple yet effective clutch. The second pulley, the one not attached to the engine, generally rests on a bearing connected to a moveable lever. The clutch lever is raised to disconnect the blades and wheel drive; the lever puts enough slack into the belt so that the engine pulley can spin freely but the belt will not rotate the second pulley.

Illustrations number 7/1, 7/2a, 7/2b and 7/3 depict the moving parts of three typical reel type lawn mower designs. Number 7/1 is the hand lawn mower now considered so old fashioned by too many people. Foot power propelled the mower and the moving wheels turned the blades via the pinion gear. In power reel mowers, the engine generally spins the reels and the pinion powers the wheels.

Illustration number 7/2a and 7/2b is an exploded view of a Yard-Man 21 inch reel mower. The clutch is activated by the mower's handle. When the handle is raised, the mower moves ahead. When the handle is dropped deliberately or accidentally, the mower wheels and blades stop although the engine itself can continue to run. Drawing 7/3 is a design by National Mower Company for both its 21 and 25 inch power mower. A manually operated clutch lever on the handle performs the same function as moving the entire Yard-Man handle.

There is at least one battery-powered electric lawn mower of the reel type on sale. The battery driven electric motor is coupled

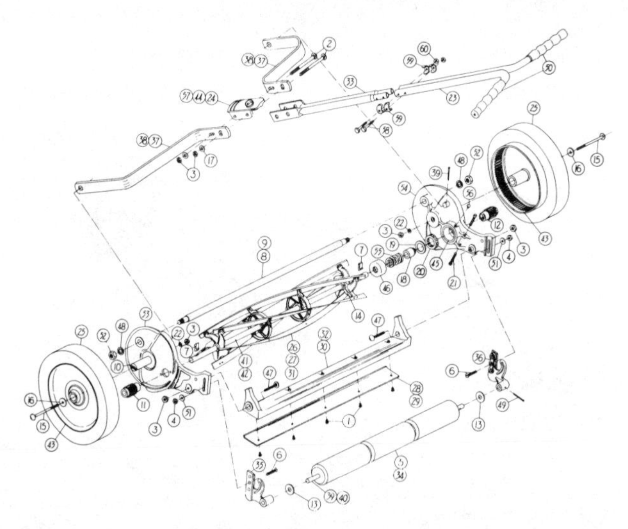

Illustration 7/1 Exploded View of Parts in a Well Built, Hand-Powered Reel Lawn Mower

Ref. No.	*Name of Part*	*Ref. No.*	*Name of Part*
1	Mach. Screw (¼-20 x ⅜ Rd. Hd.)	30	Cast Back
2	Carriage Bolt (5/16-18 x 2½)	31	Cylinder Assy.
3	Hex. Nut (5/16-18)	32	Cast Back
4	Hex. Nut (¼-20)	33	Bottom Handle Section
5	Wood Roller	34	Wood Roller
6	Mach. Screw (¼-20 x 1⅛ Flat Hd.)	35	L/H Roller Bracket
7	Pawl (⅜)	36	R/H Roller Bracket
8	Frame Shaft	37	Handle Brace
9	Frame Shaft	38	Handle Brace
10	Wheel Stud	39	Roller Shaft
11	L/H Pinion	40	Roller Shaft
12	R/H Pinion	41	Cylinder Shaft
13	Washer (29/64 ID. x 1¼ OD. x 1/16)	42	Cylinder Shaft
		43	Wheel
14	Pan Head Pin (.223″ Dia. x 1⅛)	44	Spacer Block (W/Model 2418)
15	Wheel Bolt (5/16-18 x 3¼)	45	Ball Cup
16	Wheel Washer (21/64 ID. x 1-5/16 OD. x ⅛)	46	Spring Cover
		47	Carriage Bolt (5/16-18 x 1½ Sh. Sq. Neck)
17	Washer (21/64 ID. x ¾ OD. x 1/16)		
		48	Lockwasher (½ Int. Shakeproof)
18	Bearing Cone	49	Cotter Pin (⅛ Dia. x 1½)
19	Dust Cap	50	Handle Grip (¾ Black)
20	Bearing Retainer	51	Washer (17/64 ID. x ¾ OD. x 1/16)
21	Adjusting Screw 5/16-18 x 11/8 Fillister Hd.)		
		52	Hex. Jam Nut (½-13)
22	Lockwasher (5/16 Split)	53	L/H Side Plate
23	Top Handle Section	54	R/H Side Plate
24	Spacer Block (w/Model 2416)	55	Spring
25	Tire	56	Oil Cup
26	Cylinder Assy.	57	Spacer Block (w/Model 2416-7K)
27	Cylinder Assy.	58	Hex. Cap Screw (¼-20 x 1⅞)
28	Stationary Blade	59	Special Handle Washer
29	Stationary Blade	60	Center Lock Nut (¼-20)

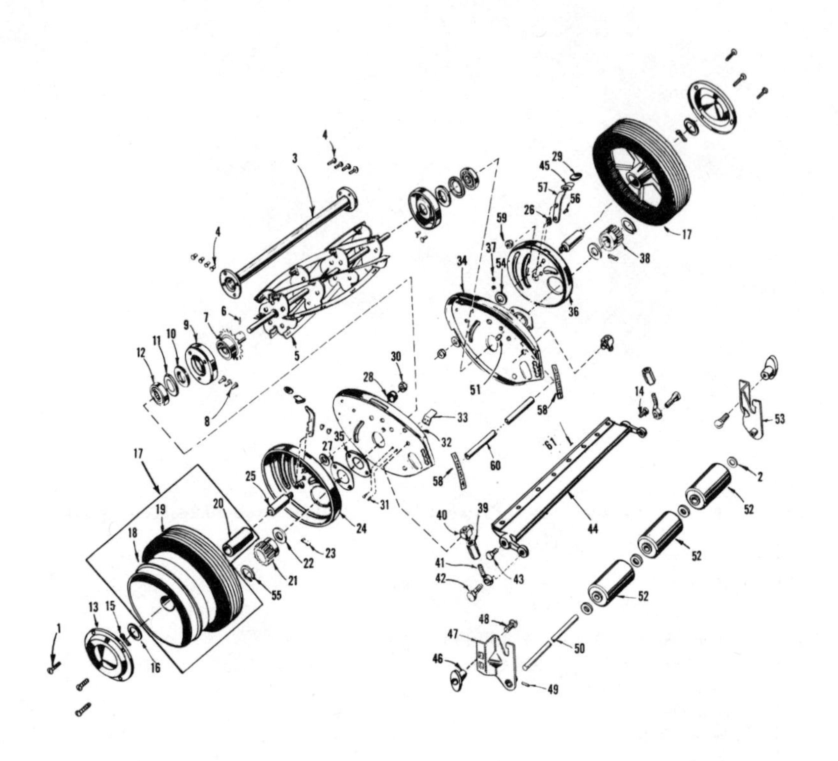
4
3
29
45
57
56
26
59
4
17
37
34
54
38
36
6
9
7
11
10
12
5
30
28
51
8
58
14
53
33
32
35
27
60
61
17
25
31
58
2
52
20
19
40
39
44
18
24
22
23
52
21
41
43
55
42
52
13
15
48
1
47
50
16
46
49

Illustration 7/2A & B A Typical Reel Lawn Mower
The clutch is engaged and disengaged by lowering or raising the handle.

Illustration 7/2A
Parts List for Illustration 7/2A

Ref. No.	Name of Part	Ref. No.	Name of Part
1	Rivet (3/16 x 7/16)	31	Screw (10-32 x 3/8 Sems)
2	Washer—Roller	32	End Plate with Trunnions—L.H.
3	Spacer Tube Assembly 21″	33	Scraper
4	Rivet (3/16 x 3/8)	34	End Plate with Trunnions—R.H.
5	Reel Assembly 21″	35	Spacer
6	Rollpin	36	Gear Housing—R.H.
7	Sprocket, Anti-wind Cup & Hub Ass'y	37	Nylon Bushing
		38	Pinion Gear—R.H.
8	Rivet (3/16 x 13/32)	39	Nut
9	Retainer	40	Trunnion
10	Felt Washer	41	Bolt
11	Felt Retainer	42	Screw
12	Bearing	43	Screw Cutter Bar—R.H. Thd.
13	Hub Cap	44	Cutter Bar Bracket 21″
14	Screw Cutter Bar L.H. Thd.	45	Dust Cover
15	Cotter Pin (1/8 x 5/8)	46	Roller Adjust Knob
16	Washer	47	Roller Bracket Assembly—L.H.
17	Wheel and Tire Assembly	48	Bolt (5/16-18 x 5/8)
18	Drive Wheel & Bearings Less Tire	49	Rollpin
19	Tire	50	Roller Shaft—21″
20	Bearing	51	Trunnion
21	Pinion Gear—L.H.	52	Roller Assembly—21″
22	Felt Washer	53	Roller Bracket Assembly—R.H.
23	Pawl	54	Washer—Special
24	Gear Housing—L.H.	55	Retaining Ring
25	Axle	56	Screw (#10-32 x 3/8 Hex. Hd.)
26	Washer	57	Spring & Trunnion Assembly
27	Retainer	58	Roller Adjust Label
28	Bushing	59	Nut (10-32 Gripco)
29	Knob Wheel Adjust	60	Spacer—End Plate (21″ only)
30	Nut	61	Cutter Bar Blade

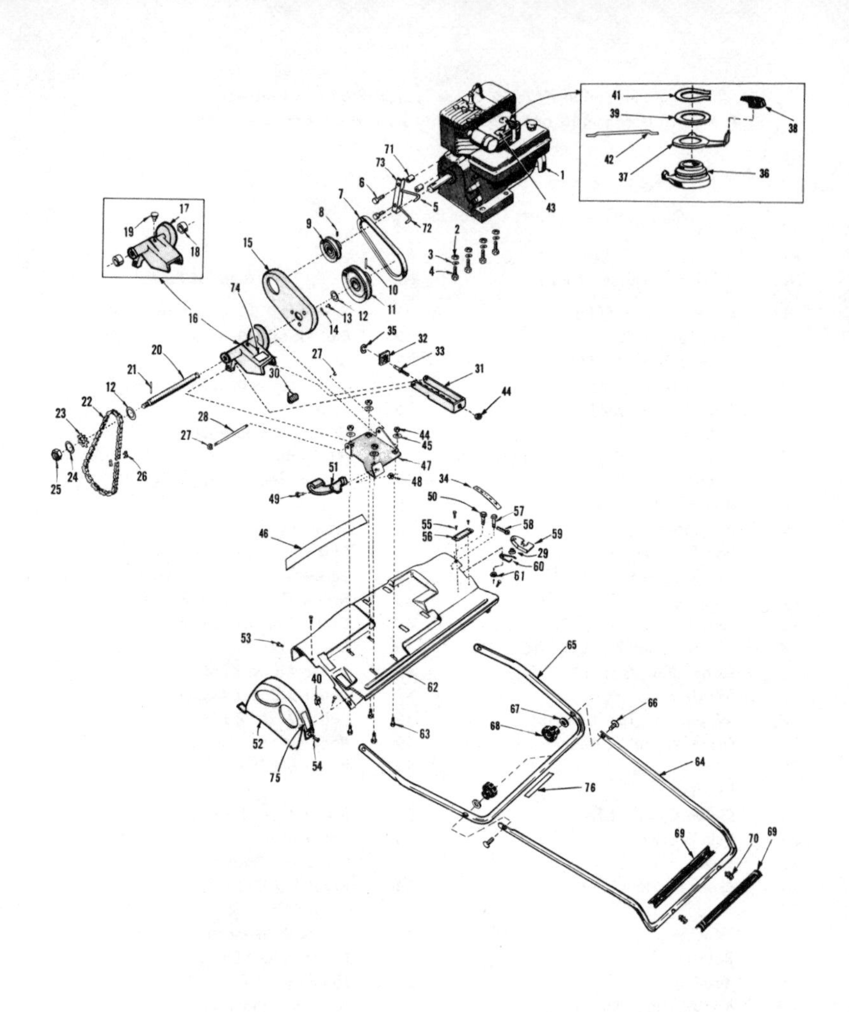
41
39
38
42
37
36
1
43
71
73
6
5
72
7
8
9
2
3
4
17
19
18
15
10
11
12
13
14
16
74
35
32
33
27
31
20
21
30
44
12
22
23
28
27
44
45
47
51
48
34
25
24
26
49
50
57
55
58
56
59
46
29
60
61
53
65
62
40
67
66
68
63
52
54
75
64
76
69
70
69

Illustration 7/2B

Parts List for Illustration 7/2B

Ref. No.	Name of Part	Ref. No.	Name of Part
1	Engine	39	Wave Washer
2	Nut (5/16-18 Hex)	40	Nut (10-32 Hex Lock)
3	Lock Washer (5/16)	41	Retaining Ring
4	Bolt (5/16-18 x 1⅜ Carriage)	42	Link—Throttle Control
5	Keeper—Belt Upper	43	Label—Throttle Control
6	Screw (¼-20 x 1¾ Fillister Hd.)	44	Nut (¼-20 Hex)
7	"V" Belt	45	Lockwasher (¼)
8	Set Screw (5/16-24 x 7/32)	46	Label—Yard-Man
9	Pulley	47	Bracket
10	Rollpin	48	Nut (5/16-24 Hex Lock Gripco)
11	Pulley Assembly	49	Bolt (5/16-24 Hex Shoulder Bolt)
12	Washer (⅝ x 15/16)	50	Bolt (5/16-24 x ¾ Hex Hd.)
13	Screw (10-32 x ½ Sems)	51	Lever
14	Washer	52	Guard
15	Belt Guard	53	Bolt (5/16-24 x ½ Hex Hd.)
16	Jackshaft Housing, Bearings & Oiler Assembly	54	Screw (10-32 x ⅜ Phlps. Hd.)
		55	Screw (6 x ¼ Rd. Hd.)
17	Jackshaft Housing	56	Name Plate
18	Bearing	57	Clevis Pin
19	Oiler	58	Cotter Pin
20	Jackshaft	59	Latch
21	Pin	60	Spring
22	Chain	61	Washer (3/16)
23	Sprocket	62	Deck
24	Washer	63	Bolt (¼-10 x ⅝)
25	Nut (7/16-20 Hex Jam)	64	Handle—Upper Section
26	Connector Link	65	Handle—Lower
27	Retaining Ring	66	Bolt (5/16-18 x 1¾ Special)
28	Pin	67	Washer—Sems
29	Spacer	68	Knob
30	Spring	69	Grip—Handle
31	Bracket—Striker	70	Pin—Grip Retaining
32	Spacer—Adjust	71	Spacer
33	Screw—Adjust	72	Lower Belt Keeper
34	Label—Wheel Adjust	73	Belt Keeper Bracket
35	Retaining Ring	74	Label
36	Adapter	75	Label
37	Lever—Throttle Control	76	Label
38	Knob—Throttle Control		

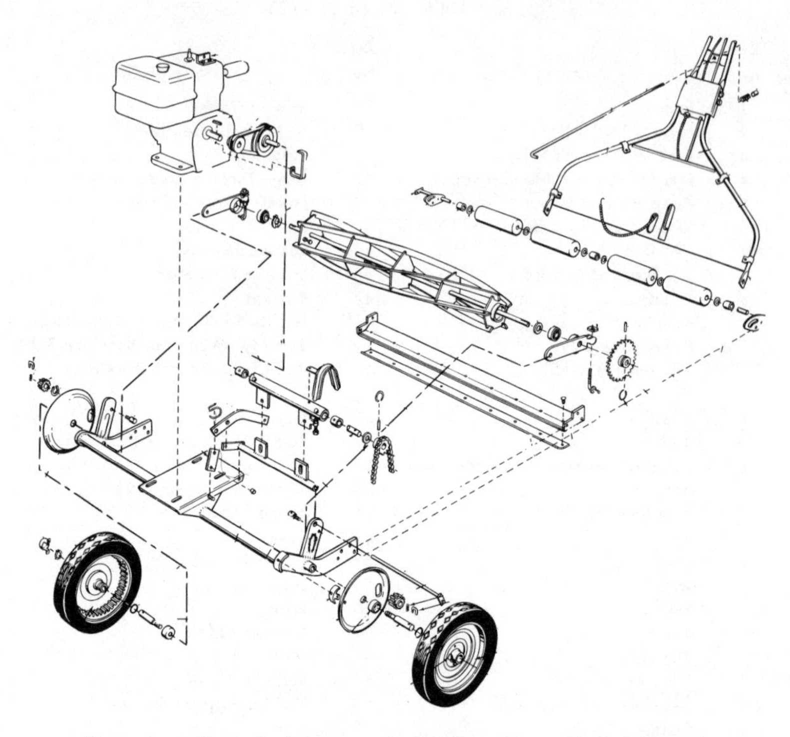

Illustration 7/3. A clutch lever on the handle activates this Reel Mower.

directly to the reel. It is not self-propelled. The operator must push not only the mower but the relatively heavy lead-acid storage battery resting on top of the blades.

ADJUSTING THE CHAIN TENSION is generally a matter of sliding the engine and the larger pulley forward or backward. Four mounting bolts typically secure the engine to the platform of a reel mower, and the holes through which the bolts pass are slotted. The larger pulley generally is held in place by two mounting bolts, also passing through slots.

Loosen the engine mounting bolts and the pulley mounting bolts. Slide the pulley assembly forward or backward to obtain the proper tension on the chain drive. The recommended setting calls for about 1/8″ of slack. Pull the pulley and sprocket assembly as tight as possible, without straining the chain, and then let it slide back approximately 1/8″. Tighten the mounting bolts and any locking nuts securely.

When properly adjusted, the chain should give noticeably when you press against it midway between the two sprockets, but it should not flop around. A chain which is adjusted too tightly tends to wear out prematurely and may break if normal or abnormal use generates increased tension against either of the sprockets. A chain which is too loose, can generate vibration as it flops around the sprockets and can gradually loosen the sprocket settings even more.

On some mowers, at least those in the Cooper Klipper line, the chain tension is adjusted with washers. Loosen the clutch plate bolts and insert enough washers of the appropriate size to tighten the chain properly.

BROKEN CHAINS ARE EASY TO REPAIR if you ever repaired your bicycle chain as a kid. In fact, should you find a bicycle shop closer than a lawn mower shop, you'll probably find all the parts needed to repair a damaged lawn mower chain.

Every chain has a *connecting link* or a *half link*. It is that one link of the chain which is designed to be removed with simple tools like a screwdriver and pliers. You can locate the removeable link by studying the side of every link in the chain as it slowly passes by. The removeable one looks obviously different; its sides are thicker and often has a broader shape. With a pliers, squeeze together the two fingers of the removeable link and then pry off the side with a screwdriver. It might spring off and fly a few inches away, so keep an eye on where it goes. With the removeable link removed, you can easily slip off the entire chain.

If one link has been damaged, examine the entire chain for other twisted or cracked links which should be repaired at the same time. To remove a damaged link from your chain, hammer out the two riveted pins which hold that particular link together. A hammer and punch is the ideal tool combination, but a nail works almost as well as a punch. Your lawn mower or bicycle store will supply you with new removeable links to replace all damaged links. Slip the two fingers of a new link into place, snap on the side plate with a pliers and make certain that the side plate is held in place by the grooves in both fingers of the link.

Lawn mower chains do not break frequently unless the chain is at least four or five years old. The most likely explanations for a break are: (1) a collision between the chain and some tough foreign object; (2) a chain which is too tight because of either poor adjustment or an accumulation of dried grass, twigs and other debris; (3) one that is badly out of line.

THE BELT DRIVE AND CLUTCH ARE ADJUSTED AFTER THE CHAIN. With the engine stopped and the spark plug wire fastened out of the way, engage the clutch so the reel would be spinning if the engine were running. Next, loosen the four engine mounting bolts but *not* the bolts which hold the larger pulley. Slide the engine and mounting bolt until the belt is tight, but don't try to stretch it. There is a certain resilience in systems of pulleys and belts which protects them from being overtightened unless you employ exceptional force. Finally, tighten all of the mounting bolts.

Disengage the clutch—move either the mower handle or the clutch lever to where the reel would not move with the engine running. With a smooth and slow action, pull the starter rope several times or activate the electric starter. Watch the pulley carefully; it should not move if the clutch is properly adjusted. (The pulley attached to the engine, of course, should move.)

If the drive pulley does move when the starter is activated, it must be adjusted to a looser position. In some mowers, you have to loosen the engine mounting bolts again and move the engine pulley and drive pulley slightly closer together. On other machines, there is a fine adjustment screw which accomplishes the same end.

Many people display a tendency to over-tighten belts and chains. Aside from the risk of breaking the chain or belt, a too-tight chain or pulley consumes a needlessly large amount of power. So, if in doubt, adjust the chain and belt just a bit toward the loose side rather than tight.

THE CUTTING BLADES HAVE BEEN THE SUBJECT OF A GOOD DEAL OF MYSTIQUE. Golf courses often have their own sharpening machine and trained technician to run it. But the average reel mower owner must take his chances on whatever commercial sharpening service is available unless he knows how to keep the blades basically in shape himself.

Once the mystique of lawn mower blades is penetrated, the rest is relatively easy. There is little reason for blades on a home lawn mower to be *sharpened* more than every two or three years if they are reasonably well cared for. *Lapping*, which some people erroneously call sharpening, but really is just polishing the knife edges, is simple enough that you can do it yourself with a screw driver and perhaps a small wrench.

Sharpening involves filing or grinding a reel blade or cutting bar to remove knicks, dents, bent knife edges or any other major injuries to the blade. Most of these imperfections are caused when a mower hits rocks, wire, bottles, toy trucks or sometimes even a heavy scrap of wood. A mower reel allowed to sit unprotected over the winter in a moist environment, will have to be sharpened to remove any severe dulling effects of rust. A third reason why blades become seriously dull is misadjustment. If the reel rubs too harshly against the cutter bar, the knife edge will be worn away, and very quickly.

Lapping accomplishes two beneficial effects. It removes small rough edges left after a reel or cutting bar has been sharpened. It also helps to align the blade and cutting bar surfaces with greater precision than a simple visual adjustment can accomplish. The better your blade alignment, the better your grass cutting.

There are two basic ways to adjust the blades on a reel type lawn mower whether it be a power mower or a hand mower. The importance of these adjustments has been over-emphasized by advertising campaigns. The *silent* mowers are really just adjusted so that the cutting bar and the reel blades do not quite touch each other. Because there is no metal-to-metal contact, there is very little noise caused by the cutting action. The gasoline engines would drown out the noise anyway so the *silent* adjustment lost its advertising usefulness when gasoline powered mowers replaced hand mowers. Now that the electric powered reel battery mower has been introduced, the *silent* setting may be back in vogue if manufacturers manage to overcome the cost and engineering headaches inherent in driving lawn mowers with batteries.

The other setting, the so-called *self-sharpening* adjustment, depends upon contact between the cutting bar and reel blades—but a *very slight* contact. The metal-to-metal friction requires somewhat more power, of course, but nothing that a 2.5 h.p. engine or a healthy adult body should notice appreciably. The cutting bar and reel blades which rub gently against each other, continuously hone the cutting edges for one another. The constant polishing action of correctly aligned cutting surfaces is very similar to lapping a set of mower blades and can help counteract dulling actions such as corrosion, rounding and curling on a knife edge. But this *self-sharpening* works only on a well cared for machine and then not indefinitely.

GRINDING a scarred or badly dulled blade is a precision skill requiring precision equipment. A really knowledgeable lawn mower blade sharpener grinds all six knife edges to within 1/64″ accuracy on his first set of passes with a grinding wheel. *And then he does it again* in exactly the opposite order to compensate for the tiny bit of unevenness which may have been introduced due to the wear his grinding wheel received during the grinding operation on your reel.

The cutting surface of a reel blade is not ground perpendicular to the blade itself but parallel to the surface of the cutting bar's knife edge. Drawing number 7/4 graphically shows that relationship. Such a design allows blades in a *silent* setting to cut as

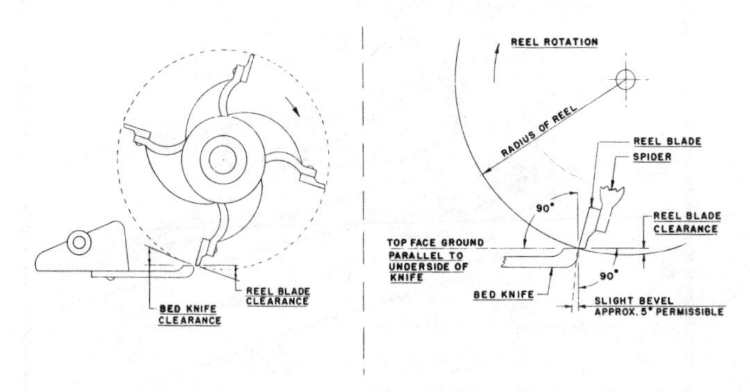

Illustration 7/4. Blueprint of Reel blade and bed knife positions.

effectively as possible without ever making contact with each other. In a *self-sharpening* adjustment, the design shown in drawing number 7/7 allows the greatest possible amount of metal-to-metal contact.

Since rotary lawn mowers dominate today's market, it is sometimes hard to find a good sharpening service for reel type lawn mowers in some localities. The dealer who sold you the mower should be your best source for the name of a qualified sharpener. But if not, try your Yellow Pages, of course, and then ask a local golf course for help. One indication of a qualified lawn mower reel sharpener is his equipment; he will have a rig designed especially for clamping a reel into place and holding it against a grinding wheel with accuracy that can be measured to 1/64". Except for major calamities, like running over a bucket of scattered bolts, there is little need to have a reel sharpened more than once every other season.

Honing—pros call it *lapping*—is required every time a reel blade or cutting bar is sharpened with a file or grinding wheel. A blade which is not badly curled or rounded by use or misuse, and has not been nicked or dented excessively, still loses its maximum sharpness due to gradual misadjustment of the blade settings and normal wear on the cutting edge. Maximum sharp-

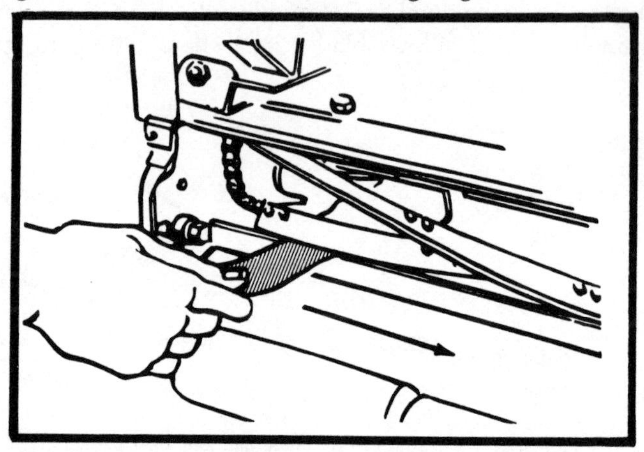

Illustration 7/5. The paper test for checking a blade's sharpness.

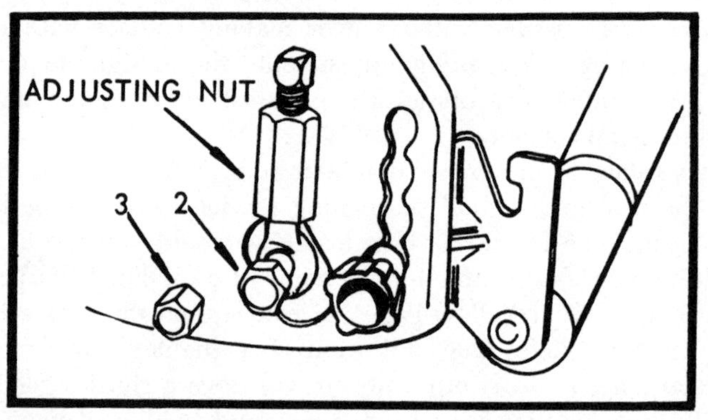

Illustration 7/6. After the blade has been correctly adjusted on a typical reel mower, tighten the set belts in this order.

ness is restored by a process called lapping which uses a suspension of fine grinding particles to give the blades a fine, smooth cutting edge.

Golf courses often lap their reels weekly. The careful lawn mower owner shouldn't have to consider lapping more than once a season, twice at the very most.

ADJUST YOUR MOWER before you begin lapping. Check all bolts and screws on or supporting the cutting bar and reel assembly. They must be in good shape and properly tightened if your adjustments are to last. Make certain that the reel bearings are sound; if not, adjust them or replace them. Repack the bearing with grease unless instructions for your mower specify otherwise.

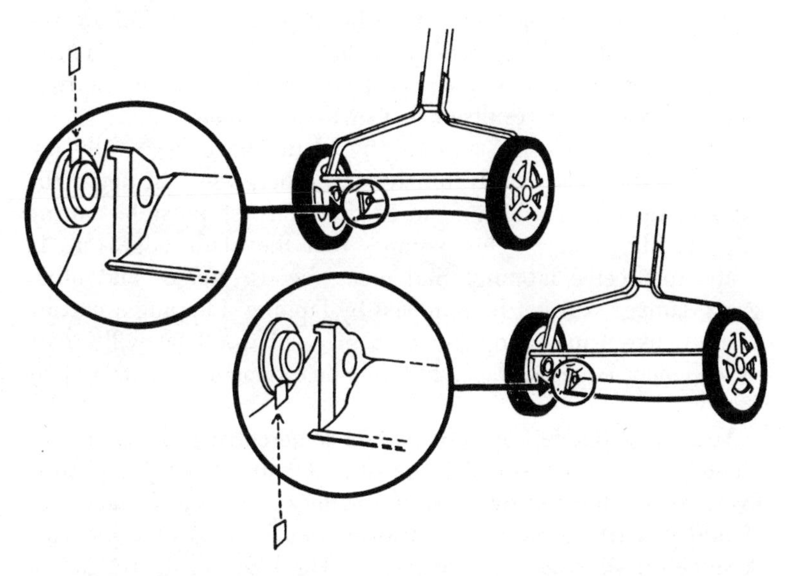

Illustration 7/7. How to put distorted mowers back in line by shimming bed knives with paper.

Rotate the reel slowly, studying how each knife blade contacts or almost contacts the cutter bar along its entire length. Insert small slips of newspaper to help evaluate whether the reel blades touch the cutter bar evenly at every point. Each of the five, six or seven blades must contact the cutter bar identically. If the blades touch harder at the right side then the left, for example the right side must be adjusted to reduce the amount of contact, and vice versa.

If the blades rub gently on the extreme right and left portions but hard in the center, the cutter bar appears to be bowed upward. Such a condition often can be corrected by inserting a shim above the side plate which supports the cutter bar. Illustration number 7/7a demonstrates how. And for the reverse condition, gentle contact in the center of the cutting bar but too much contact on either end, the bar appears to be bowed downward and should be corrected according to the suggestion in illustration number 7/7b.

Adjustments to the cutting surfaces, either the reel or the cutting bar, should be made very slowly. One-eighth of a turn on the adjusting screws is the most you should make at one time before checking the results of that small adjustment.

After the cutting bar and cutting blades have been fully adjusted so they just barely touch each other, you can begin the lapping operation which restores the sharpest possible cutting edge to all surfaces. This assumes that the blade edges are in shape to receive lapping. Bad nicks, bends, gouges and other gross damage will not be removed by lapping. Lapping a cutting blade is like using sandpaper on a piece of wood—it will create or enhance a fine surface, but it won't remove serious imperfections.

You might ignore a nick or two on some cutting blades. If you want to maintain the reel in its utmost cutting condition, however, you should not overlook too many or too heavy scars. Nor should you try to file or grind away visual traces of a blemish. A stone on your lawn might have left a 1/32″ or 1/16″ gap in the cutting edge on one of your blades, but if you grind on only that single portion of one blade, you will actually increase the amount of unusable cutting surface.

LAPPING IS GENERALLY ACCOMPLISHED BY RUNNING A MOWER REEL BACKWARDS. A mild grinding material, called lapping compound, polishes the cutting surfaces enough to renew their former keen edge. The Black & Decker battery powered reel mower allows its user to reverse the polarity of the motor. The reel then moves backward under electric power for the lapping operation. Gasoline engine powered models or hand powered models require some other, less simple arrangement.

Commercial operators can afford to buy a lapping device which consists of little more than an electric motor with fittings which enable the motor to spin a lawn mower reel backward. An ingenious home handyman might rig up a similar device for his hand or gasoline powered mower.

Since the typical power reel mower is belt driven, if the belt were twisted, the reels would revolve backwards during a lapping operation. Success in accomplishing this depends in part upon the construction of your own individual lawn mower. If the distance between pulleys is long enough or the belt adjusting slots generous enough, twisting the belt would be relatively simple. If not, you might experiment with a slightly oversized belt.

Depending upon the specific construction of the pinions and pawls in the wheel assembly of your mower, the drive wheels may or may not be driven. If they do revolve during lapping, they must be blocked off the ground. Test this carefully in advance of starting the engine by first rotating the reel backwards by hand to see if the wheels follow suit. Do not run the engine at a very high speed during your lapping operation, probably little above idle.

In most localities, an automotive supply store is the quickest source of lapping compound. It is sold for use in valve grinding and valve lapping operations. Other grinding compounds are available, often in hardware stores. Specify an oil mixed sharpening compound with grit sizes from 100 to 300 microns. A water mixed compound can be used if an oil mixed one is hard to find.

When your lapping operation is ready to begin, apply lapping compound to one or more of the blades. If the compound is fairly thick, an application to only one blade may suffice because the mixture will be dispersed at the cutting bar. Should the compound be much thinner than the consistency of toothpaste, apply it to all of the reel blades, and frequently.

How long this lapping should continue depends on the lapping compound, the speed at which the knives are rotated and how dull or how badly nicked they were when you began. Ten minutes to half an hour are logical lower and upper time limits for your first try at lapping. An hour may be needed to restore badly dulled blades. On blades which began with considerable roughness on their cutting edges, it may be necessary to readjust the spacing between reel and cutting bar midway through the lapping operation, by tightening the adjustment by no more than 1/8 of a turn.

After your blades have been successfully lapped back into prime cutting condition, carefully adjust the relative position of the cutting surfaces as described earlier. Don't forget to untwist the pulley after you're done lapping. The reel blades should again just barely touch or just barely miss touching the cutting bar evenly across the entire length of the cutting bar. You can even test the blades for sharpness with a few strips of newspaper as demonstrated in illustration number 7/5.

For the *silent* setting, where blades come very close but never actually touch each other, a strip of newspaper should be crisply pinched or better still, folded, by the action of the reel blade and the cutting bar. For a *self-sharpening* adjustment, the paper should be cleanly sliced. Run this newspaper test at three or four different points along the length of the cutting bar.

After you've honed your own instincts enough to get your two or three year old reel mower passing the newspaper test, you know that you've made it. The rest of the 10 years or more that your mower ought to last will be all down hill!

CHAPTER 8

Rider Lawn Mower Repairs

RIDING lawn mowers come in an unbelievable number of sizes, shapes, powers and designs. The functions which the various mechanisms must perform in a riding mower are similar, but the actual equipment can be in the rear, in the front, on top, underneath or built into other sub-mechanisms.

Essentially a riding lawn mower needs a frame, a seat, and an engine which has to drive both the wheels and mower blades, a feat which it does with either pulleys or gears, often with both. And there has to be some way to steer.

THE VARIATOR MECHANISM in drawing number 8/1 shows one way that engineers use pulleys to bring power from the engine up front to the drive wheels in back by way of both a clutch and a variable speed device. Variations on this type of system are found in several riding lawn mowers.

The engine in the design shown in drawing number 8/1 drives the mower blades with its main crankshaft. The power take-off section (P.T.O.) of the engine powers the wheels via the various sized pulley sheaves. The belt connected to the engine's P.T.O. spins the variator which is a set of two coupled pulleys whose relative size can be varied by moving the variable speed control lever. In the fastest setting, the sheaves for the engine belt form as small a pulley as possible and the sheaves for the drive wheel belt form as large a pulley as possible. At the slowest setting the relative sizes are reversed.

That pulley which looks like an extra thumb, the *idler*, is more

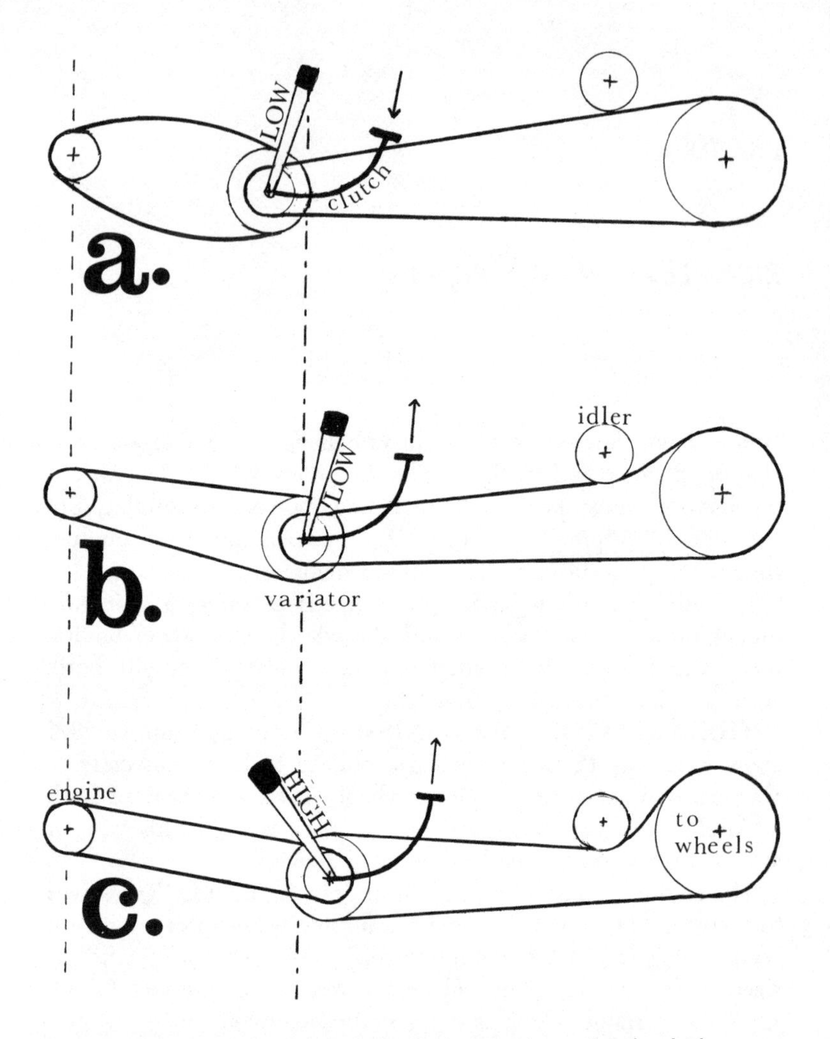

Illustration 8/1. Variator. System of speed selection, Disengaging the clutch removes tension from the pulley between engine and variator (a), allowing the engine to spin but not transmitting the motion to wheels. In its low-speed positions (b), the variator couples to the engine via a large pulley, and couples to the wheels via a small pulley. And for high speeds (c), the engine belt drives a small pulley in the variator, and the wheel belt is driven by a large pulley sheave.

useful than it might seem at first. The drive wheel belt is several inches too big so the variator is able to move about. The idler pulley keeps tension on the secondary pulley by taking up the slack so the secondary belt can transmit its power to the drive pulley. When tension is taken off the idler pulley, the secondary belt is very slack so the engine keeps on running but the drive wheels will not move forward or backward. In short, the idler can act as a simple clutch also.

Most mower manufacturers using a variator or similar belt driven system do not, in fact, use the idler pulley as a clutch. They prefer to incorporate a clutch which creates slack in the belt from the engine rather than in the belt going to the drive wheels. In the John Deere variator system, for example, the clutch is attached to the variator pulley itself, in a manner similar to drawing 8/1.

Belts in all belt-driven mowers must be kept in good condition and replaced whenever wear, breaks or stretching is noticed. Keep oil and grease away from belts; they not only reduce the power which a belt can transmit, they deteriorate the rubber chemically. You should adjust mower belts in accordance with whatever user manual may have come with your mower. In general, a belt with an idler system on it should be kept tight enough so that, when the idler is taking up the belt slack, you cannot push the belt more than an inch out of line midway between pulleys if you push with only one finger.

Pay particular attention to user manual instructions about adjusting the power take-off clutch or brake. That is the device which stops the whirling blades, hopefully before you leap off the mower and perhaps accidentally shove your hand or foot underneath the mower and near a rotary blade. Two drawings in the rotary mower section (illustrations number 6/8 and 6/9) show brake arrangements and adjustment. Without a properly adjusted power take-off or blade brake, the blade or blades can keep whirling for 15 or 20 seconds after you've stopped the mower, maybe more.

Variable pulley sheaves create an efficient means of regulating the forward speed of lawn mowers. John Deere reports that its variator system varies the forward speed from 0.4 to 7.4 m.p.h. All of the belts and pulleys used in such a system must be kept thoroughly clean and well adjusted. The variator must be able to move through several positions, each creating different tensions on the engine belt and drive wheel belt, while not interfering with the clutch's ability to disengage the variable sheaves. In making adjustments, do not run the engine. Check for proper performance while the clutch is alternately engaged and disengaged and while the variator is set in high, low and medium settings.

RUBBER DRIVE SYSTEMS. Another deceptively simple speed regulator is found in riding mowers like Lawn Boy, Poloron and Ariens. These mowers use wheels turning at right angles to each other both as a speed control and a clutch. The *friction wheel* spins in a fixed position and generally at a fixed speed; it is aluminum in a Lawn Boy rider, rubber in Ariens and Poloron. Rubbing against the friction wheel is the *power wheel,* made of rubber. The power wheel can move toward the center or outside of the friction wheel. At the outside, the speed is greatest. Toward the center the speed is slowest. And at the very center, the power wheel does not move at all—it's in neutral, you could say. Lawn Boy chooses to pull the drive wheel 1/16″ away from the friction wheel in "neutral."

A friction system based on rubbing rubber wheels would never work efficiently at the high speeds you enjoy in your car. But for the low-power, low-speed service required in riding lawn mowers, the rubber wheels work fine. Drawing number 8/2 illustrates the principle. This assumes, of course, that you are able to keep the adjustment close to par and replace the rubber friction and drive wheels as needed. Carefully lubricate all moving parts in the vicinity of friction power control systems because oil or grease can thoroughly foul them up.

DIFFERENTIALS ARE FOUND ON THE REAR WHEELS OF LARGER OR BETTER RIDING MOWERS. With front-wheel steering, whenever you turn a corner the rear wheels must go at different speeds. On a sharp corner, the inside wheel might scarcely turn at all while the outside wheel could go around

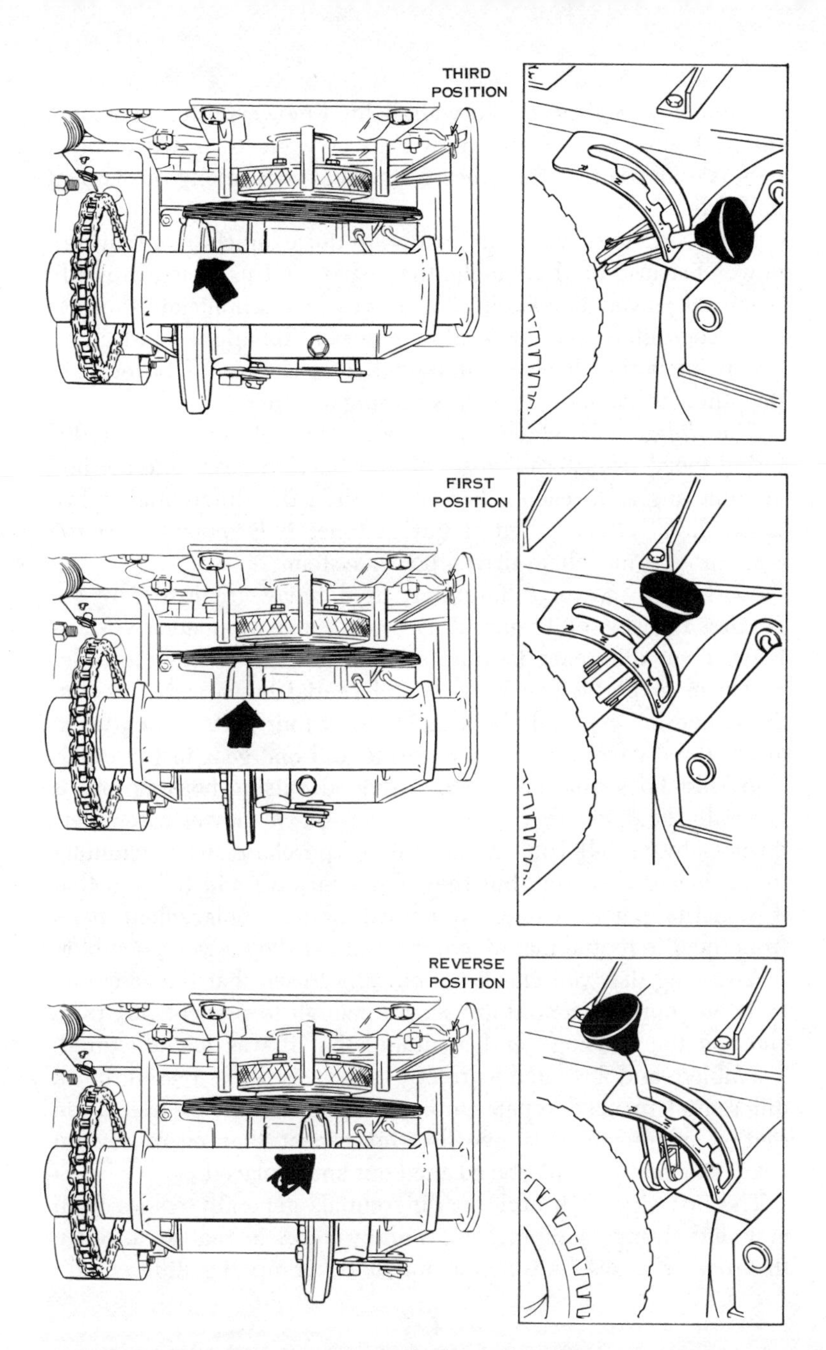

Illustration 8/2. A revolving wheel system of power Transmission. The friction wheel (power wheel) moves at a fixed speed. The drive wheel rubs against the outside left edge for a "high-speed" output and near the inside for "low-speed." It moves to the right side for "Reverse."

continuously. Self-propelled walk-behind mowers often overcome this difficulty by applying power to only one rear wheel on a half-axle and letting the other rear wheel coast along on its own half-axle.

Since riding mowers are wider, require much more overall power because of the weight they carry and need more evenly balanced power than walkers, a set of gears is built into the rear axle. The differential gears allow power to be applied just about evenly to both wheels even though one is turning faster than the other, as when the mower is turning a corner.

The *differential* which allows power wheels to move at differing speeds is generally a set of four beveled gears intermeshed at right angles to each other. Very often the differential is fastened directly to one wheel but at times it is mounted on the rear sprocket in a chain-driven power system.

Most lawn mower differentials used today are made by the Peerless Division of Tecumseh Engines. Some are made by Mast-Foos. Their difference is quite esoteric unless you have to order replacement parts. The Peerless differential has a smooth cylindrical housing generally held together by four bolts. When taken apart, three gears remain in one side and one gear in the other. The Mast-Foos differential has a shaped outside housing and is generally held together with six bolts. Lawn mower assemblers do not always tell you, at the time of purchase, who manufactured their differential, but they almost always will tell you that if something goes wrong, you must obtain replacement parts from the differential manufacturers and not the mower assembler.

Assuming that you check periodically to see that the assembly bolts on your differential are secure enough to prevent dirt from entering the housing—and assuming that it was manufactured, assembled and installed correctly—there is little reason for a differential to need repair until it has seen a lot of use. Even routine maintenance is uncalled for except that every several years, the grease should be cleaned out and replaced.

The first sign of trouble in differentials generally comes from new and strange scraping or grinding noises in that area. Since the roar of a gasoline engine might cover up the differential's first cry for help, your second warning probably will come with jerky performance in the drive wheels. The mower blades themselves will jerk if the engine is at fault. If the wheels jerk but not the blades, suspect your differential. Then you will have to

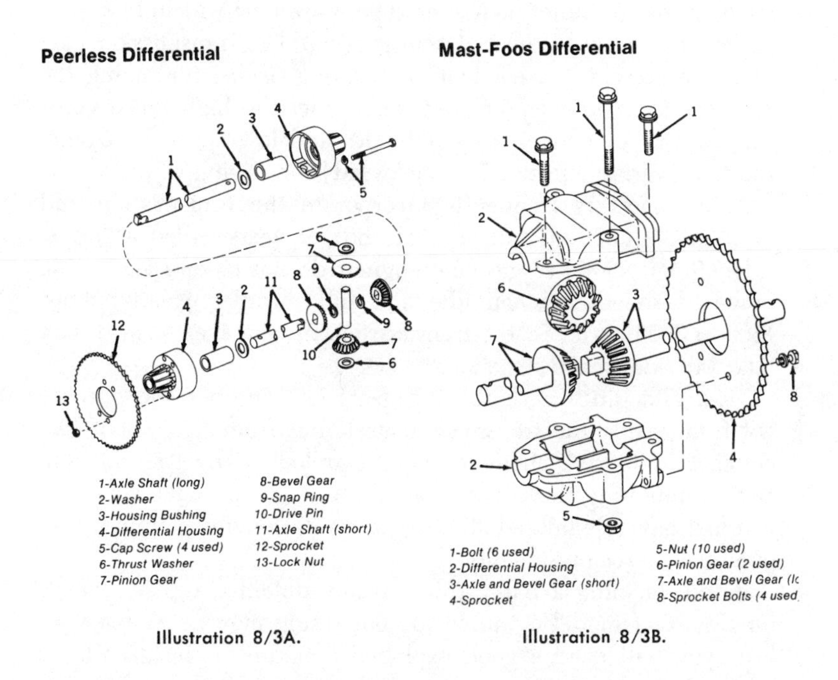

Illustration 8/3A.

Illustration 8/3B.

remove the differential, probably by first removing the appropriate wheel, and disassembling it to find out if the gears are worn or broken. Perhaps dirt has worked its way inside the housing.

A differential is too expensive to replace with an entirely new one at the first sign of trouble. You will find many dollars worth of incentive in studying the drawings numbered 8/3a and 8/3b to identify the names for whichever parts appear to need replacing. Every time a differential is taken apart, the grease inside should be cleaned away and replaced with fresh EP grade lithium lubricant.

TRANSMISSIONS FULL OF GEARS are found on many mowers and serve to regulate speed. At the very least, the forward and backward direction generally is produced by a simple set of gears. At times, a tractor type mower will include *both* a pulley system and a geared transmission. The transmission gear system represents an efficient engineering device for taking the power of a gasoline engine and converting it to high speed vertical rotation for driving mower blades and low speed horizontal rotation (variable forward or backward) for driving wheels.

Riding mowers frequently incorporate the transmission and differential functions into a single box of gears called a "transaxle." Single cylinder gasoline engines do not have a very wide range of speed between idle and full throttle, usually about 1500 to 3000 r.p.m. So the transmission, in whatever form it may take, provides speed selection.

Most difficulties with shifting gears or improper speed come not from inside the transmission itself but from equipment like clutches and drive chains or belts coupled to the transmission. Before unbolting the cover of your transmission, make sure that you first have eliminated all other items in the power train as the potential source of difficulty.

Before tackling a balky transmission, obtain a repair manual for the exact model mounted to your riding mower. At the very least you will need a good exploded diagram of the gear locations. To tackle a transmission without a manual or exploded diagram, you must be either very nervy or very rich. Or a highly systematic person who can take out each gear in its turn, examine it, clean it and put it onto a workbench in such an organized way that you'll be able to shove the gears back together again in the same order they came out.

A great many transmissions and transaxles found on riding mowers again are made by the Peerless Division. You may not be aware of this fact until your lawn mower dealer tells you to go find a Tecumseh dealer if something goes wrong with the set of gears strapped onto your lawn mower. John Deere and Yardman use identical transmissions on some of their riding equipment, but since they supply parts and diagrams themselves, it probably isn't made by Peerless.

The insides of typical transmissions and transaxles are illustrated by several representative exploded drawings near by (numbers 8/4, 8/5, 8/6 and 8/7).

Peerless recommends *S.A.E. 90 extra duty* transmission oil for their transaxle and transmission. The transmission oil level should be checked frequently and brought up to the proper height if needed. Old oil must be drained and replaced with fresh oil after every 50 hours of operation—more frequently in dusty weather.

The transmission included in the Deere riding mower and the Yardman 3420-1 uses a lithium grease which fills the entire bottom half of the transmission housing. You can be reasonably certain that any riding mower sold with what is often called a "sealed for life" transmission is filled with grease instead of oil. Grease will not ooze out of a small leak like oil, so even if the transmission housing warps or the gasket rots away, the gears will still be inside a lubricated environment.

A weekend grass cutter accustomed to driving a car with automatic transmission, slipping from *forward* to *reverse* and back to *forward* without hesitation, can ruin the simpler riding lawn mower transmission. Most of the geared transmissions *cannot* be shifted while the mower is in motion. At least they *should not* be. If you work hard enough, you'll probably be able to shift while rolling across the turf, but you'll be damaging the transmission at the same time.

The most frequent malfunction within a transmission housing is caused by bent or broken shift forks and whatever rods or levers attach to them. The shift forks are what actually shift the position of some of the gears backward and forward when you move the shifting rod. With a certain amount of luck and persistence, you should be able to spot damaged shift forks without having to pull out every last one of the gears inside a transmission.

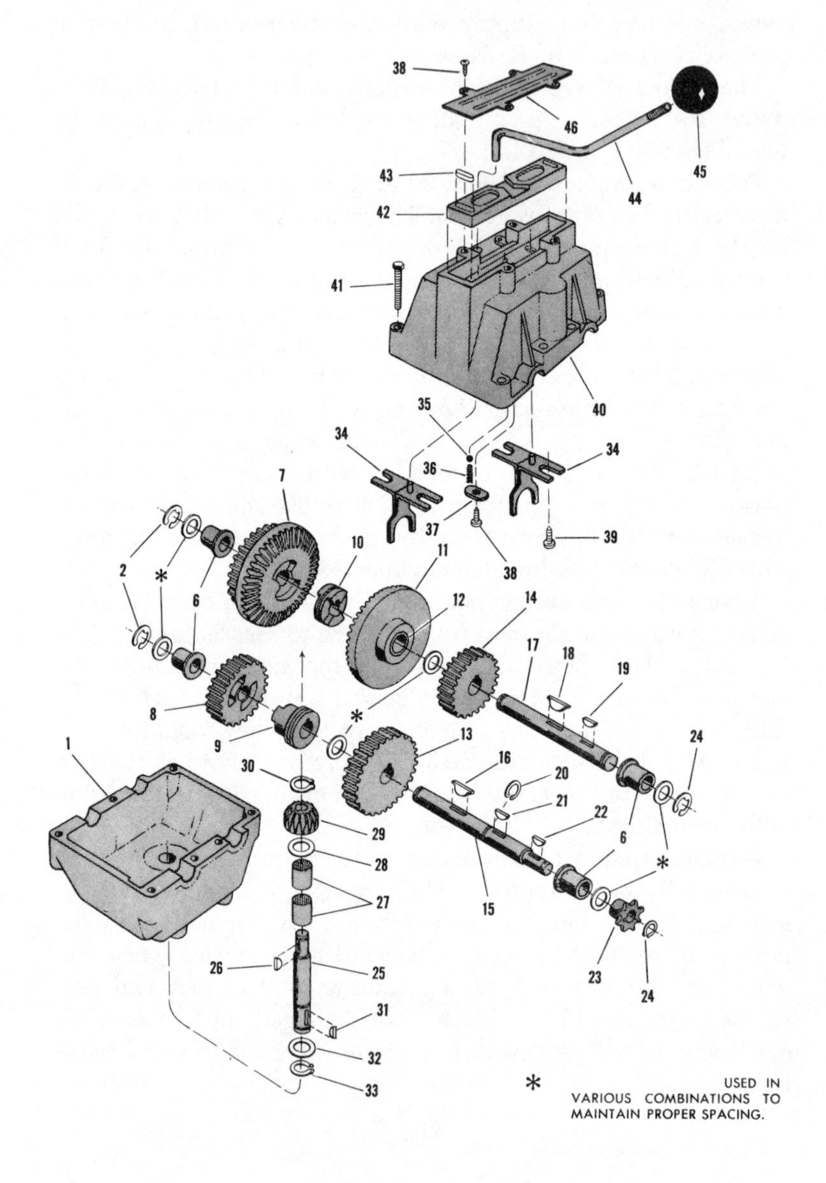
38
46
45
43
44
42
41
40
35
34
34
36
7
37
39
10
38
11
2
*
6
12
14
17
18
19
8
*
9
13
24
1
30
16
20
21
22
29
6
28
*
27
15
23
26
25
24
31
32
33
* USED IN VARIOUS COMBINATIONS TO MAINTAIN PROPER SPACING.

Illustration 8/4 Two-Speed Transmission: Parts List
(Your Mower is right hand (R.H.) or left hand (L.H.) as you mow.)

Ref. No.	Name of Part	Ref. No.	Name of Part
1	Housing—Lower	24	Ring—Retaining
2	Ring—Retaining	25	Shaft—Input
3	Washer (.040)	26	Key—Woodruff #2
4	Washer (.030)	27	Bearing—Needle
5	Washer (.050)	28	Washer—Thrust
6	Bearing—Flanged	29	Gear—Bevel, Pinion
7	Gear—Spur & Bevel	30	Ring—Retaining
8	Gear—Spur (12T without keyway)	31	Key—Woodruff #6
9	Collar—Clutch	32	Washer—Thrust
10	Collar—Clutch	33	Ring—Retaining
11	Gear—Bevel (42T)	34	Fork—Shifter
12	Bearing—Sleeve	35	Ball—Detent
13	Gear—Spur (28T)	36	Spring—Detent
14	Gear—Spur (22T)	37	Plate—Detent Cover
15	Shaft—Output	38	Screw (8-32 x ⅜ truss hd.)
16	Key—Special Hi-Pro	39	Screw—Shoulder
17	Shaft—Drive	40	Housing—Upper
18	Key—Special Hi-Pro	41	Bolt (¼-20 x 1-5/16 hex. hd.)
19	Key—Woodruff #3	42	Plate—Cam
20	Ring—Retaining	43	Slide—Nylon
21	Key—Woodruff #61	44	Lever—Shift
22	Key—Woodruff #4	45	Knob—Shift
23	Sprocket—8T #41	46	Plate—Cover Top

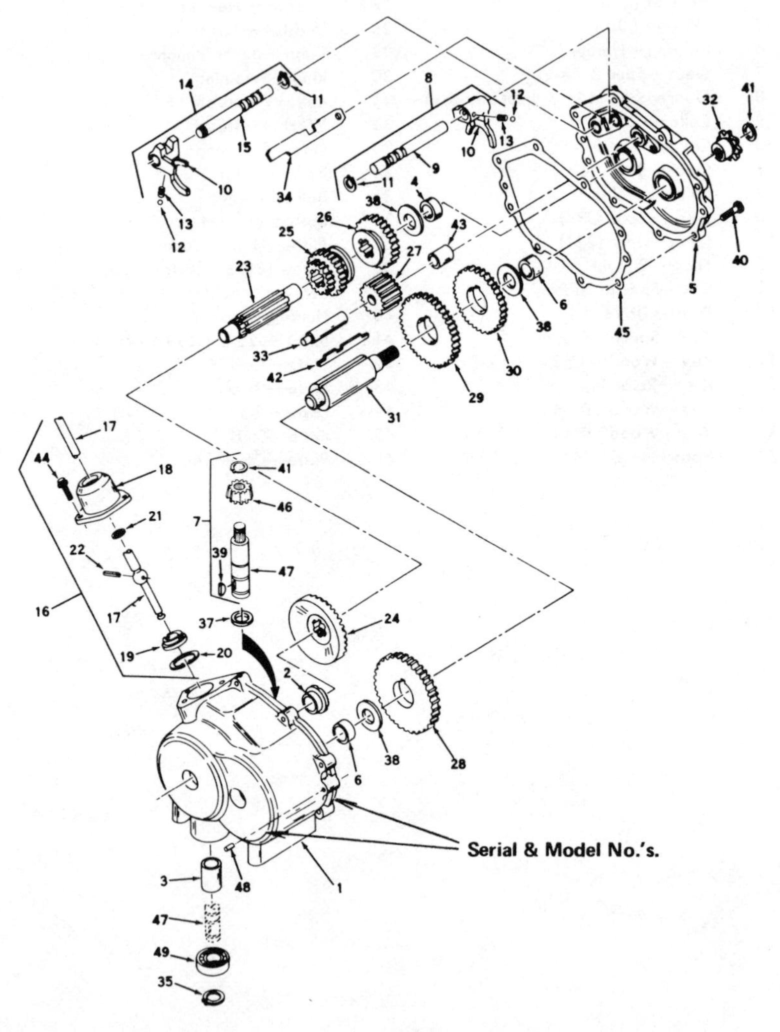
Serial & Model No.'s.

Illustration 8/5 Tecumseh Model No. 353 Transmission: Parts List

Ref. No.	Name of Part
1	Case Assy., Transmission (Incl. Nos. 2, 3 & 6)
2	Bearing, Bronze
3	Bearing, Bronze
4	Bearing, Bronze
5	Cover Assy., Transmission (Incl. Nos. 4 & 6)
6	Bushing, Bronze
7	Shaft & Gear Assy., Input (Incl. Nos. 41, 46 & 47)
8	Rod Assy., Shift (Incl. Nos. 9 thru 13)
9	Rod, Shift
10	Fork, Shifter
11	Ring, Retainer
12	Ball, Steel
13	Spring
14	Rod Assy., Shift (Incl. Nos. 10 thru 13 & 15)
15	Rod, Shift
16	Lever & Housing Assy., Shift (Incl. Nos. 17 thru 22)
17	Lever, Shift
18	Housing, Shift Lever
19	Keeper, Shift Lever
20	Ring, Snap
21	Ring, Quad
22	Pin, Roll 3/16 x 15/16
23	Shaft, Shifter
24	Gear, Bevel (33 teeth)
25	Gear, Shift (1st & 2nd)
26	Gear, Shift (3rd & Rev.)
27	Idler, Reverse
28	Gear (39 teeth)
29	Gear (34 teeth)
30	Gear (30 teeth)
31	Shaft, Output
32	Sprocket (8 teeth)
33	Shaft, Reverse Idler
34	Stop, Shifter
35	Ring, Retainer
37	Bearing, Thrust
38	Washer
39	Key, Woodruff No. 61
40	Screw, Hex. hd. self tap ¼-20 x 1
41	Ring, Retainer
42	Key, Counter Shaft
43	Spacer, Reverse Idler
44	Screw, Socket hd. cap ¼-20 x ¾
45	Gasket, Transmission
46	Pinion, Input
47	Shaft, Input
48	Pin, Dowel
49	Bearing, Ball

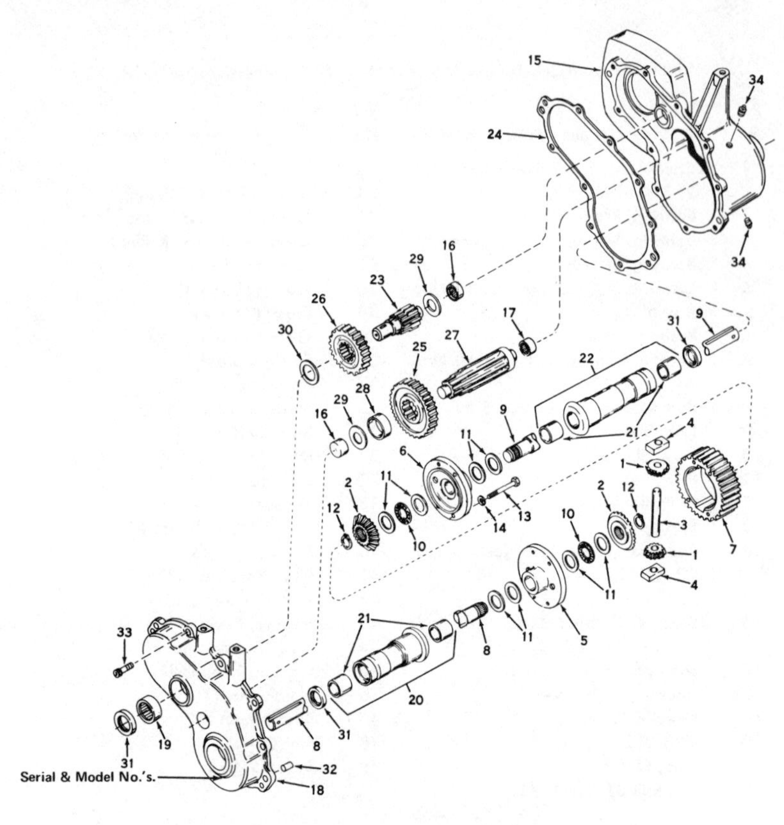

Illustration 8/6 Tecumseh Model 1303 Transaxle: Parts List

Ref. No.	*Name of Part*
1	Pinion, Bevel
2	Gear, Bevel
3	Pin, Drive
4	Block, Drive
5	Carrier, Differential
6	Carrier, Differential
7	Gear, Ring
8	Axle, L.H.
9	Axle, R.H.
10	Bearing, Thrust
11	Washer, Thrust
12	Ring, Snap
13	Screw, ¼-20 x 2¼ Hex. hd. cap
14	Lockwasher, ¼″
15	Case Assy. (Incl. No's. 16 & 17)
16	Bearing, Needle
17	Bearing, Needle
18	Cover Assy. (Incl. No's. 16 & 19)
19	Bearing, Needle
20	Housing & Bushing Assy., L.H. (Incl. No. 21)
21	Bushing
22	Housing & Bushing Assy., R.H. (Incl. No. 21)
23	Shaft, Brake
24	Gasket, Case to cover
25	Gear, Output
26	Gear, Idler
27	Shaft, Output
28	Spacer
29	Washer
30	Washer
31	Seal, Oil
32	Pin, Dowel
33	Screw, ¼-20 x 1 Hex. hd. self tapping
34	Plug, Pipe

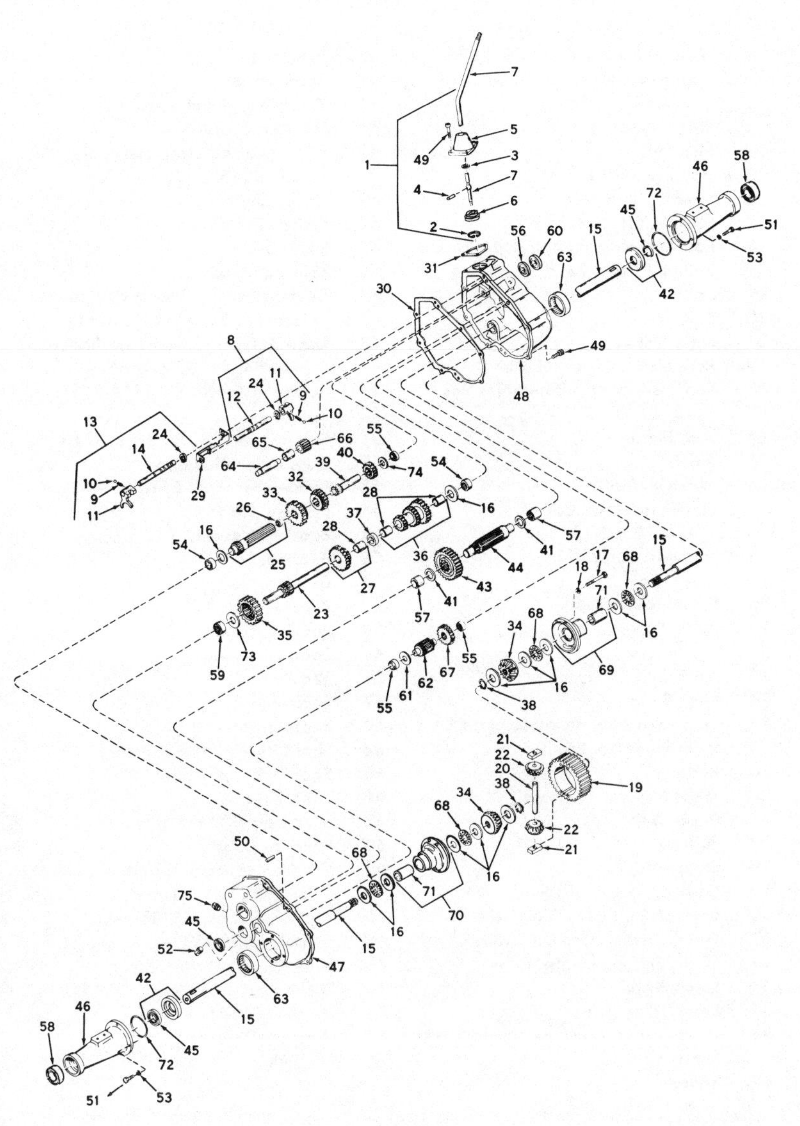

Illustration 8/7 Exploded View of a Transaxle

Parts List

Ref. No.	Name of Part	Ref. No.	Name of Part
1	Lever and Housing Assembly, (Includes Numbers 2 thru 7)	38	Ring, Snap
		39	Shaft, Input
2	Ring, Snap	40	Spur Gear, Input Shaft
3	Ring, Quad	41	Washer, Thrust
4	Pin, Roll	42	Seal and Retainer Assembly, Oil (Includes No. 45)
5	Housing, Shift Lever		
6	Keeper, Shift Lever	43	Gear, Output
7	Lever, Shift	44	Pinion, Output
8	Rod Assembly, Shift (Includes Nos. 9 thru 12 and 24)	45	Seal, Oil
		46	Housing, Axle
9	Spring	47	Cover Assembly, Transaxle (Includes Nos. 54, 55, 57, 59 and 63)
10	Ball, Steel		
11	Fork, Shifter	48	Case Assembly, Transaxle (Includes Nos. 54, 55, 57 and 63)
12	Rod, Shifter (3rd and 4th)		
13	Rod Assembly, Shift (Includes Nos. 9, 10, 11 and 14)	49	Screw, Socket Hd. Cap 1/4-20 x 3/4
		50	Pin, Dowel
14	Rod, Shifter (low)	51	Screw, Hex. Hd. 5/16-18 x 7/8
15	Axle	52	Plug, Magnetic Drain
16	Washer, Thrust	53	Lockwasher, Split 5/16″
17	Screw, Hex. Hd. Cap 1/4-20 x 2-1/2	54	Bearing, Needle
		55	Bearing, Needle
18	Lockwasher, 1/2″	56	Bearing, Ball
19	Gear, Ring	57	Bearing, Needle
20	Pin, Drive	58	Bearing, Ball
21	Block, Drive	59	Bearing, Needle
22	Pinion, Bevel	60	Seal, Oil
23	Shaft and Gear, Brake	61	Washer
24	Ring, Snap	62	Shaft and Pinion
25	Shaft and Bearing Assembly, Piston (Includes No. 26)	63	Bearing, Needle
		64	Shaft, Reverse Idler
26	Bearing	65	Spacer, Reverse Idler
27	Gear Cluster Assembly (Includes No. 28)	66	Idler, Reverse
		67	Spur Gear (22 teeth)
28	Bushing	68	Bearing, Thrust
29	Stop, Shifter	69	Carrier Assembly, Differential (Includes No. 71)
30	Gasket, Case and Cover		
31	Gasket, Shift Lever Housing	70	Carrier Assembly, Differential (Includes No. 71)
32	Gear, Shifting (3rd and 4th)		
33	Gear, Shifting (3rd and 4th)	71	Bushing
34	Gear, Bevel	72	"O" Ring
35	Gear, Idler	73	Washer, Thrust
36	Gear Cluster Assembly, Oil (Includes No. 45)	74	Washer, Thrust
		75	Plug, Pipe
37	Spacer		

While inspecting a transmission, watch for gears with broken or worn teeth. If you find one that's worn, look further; you'll probably find more. A single worn gear out of the dozens inside a transmission probably indicates a worn bushing or bearing connected to that particular gear, assuming that everything was properly assembled at the factory or reassembled properly after some earlier disassembly.

STEERING MECHANISMS require a great deal more care than you would imagine, judging by the amount of space which user manuals devote to them. It is a rare manual which includes more thon a line or two about the total steering system. It is an even rarer manual which includes a comprehensive drawing of the steering assembly.

Frankly, the entire progression of moving parts linking the steering wheel to the wheels all too often is held together by a small bolt or two and a couple of snap-rings. That might account for the assemblers' apparent reluctance to blueprint their steering systems. Despite the fact that well-greased wheel pivots are necessary to insure smooth steering, some riding mowers have no grease cups! Their manufacturers recommend that you squirt light oil on top and hope that some of it seeps down into the mechanism.

Drawing number 8/8 outlines a typical steering layout for a riding mower. The wheels themselves are held onto the *axle* by the *spindles,* which look like bent iron rods with a small iron *arm* welded on. The wheel is free to spin on one end of the spindle and the other end of the spindle moves freely about inside the axle. *Tie rods* link each arm of the *steering arm.* The steering arm in turn is either attached directly to the *steering column* or is indirectly linked to the column by a set of gears. And the steering column, of course, ends with the familiar steering wheel on top.

Finer steering systems have refinements added to just about every step in the simple set up above. There can be a seal and bearing where the steering column passes through the frame, for instance. There may also be ball bearings at the bottom as

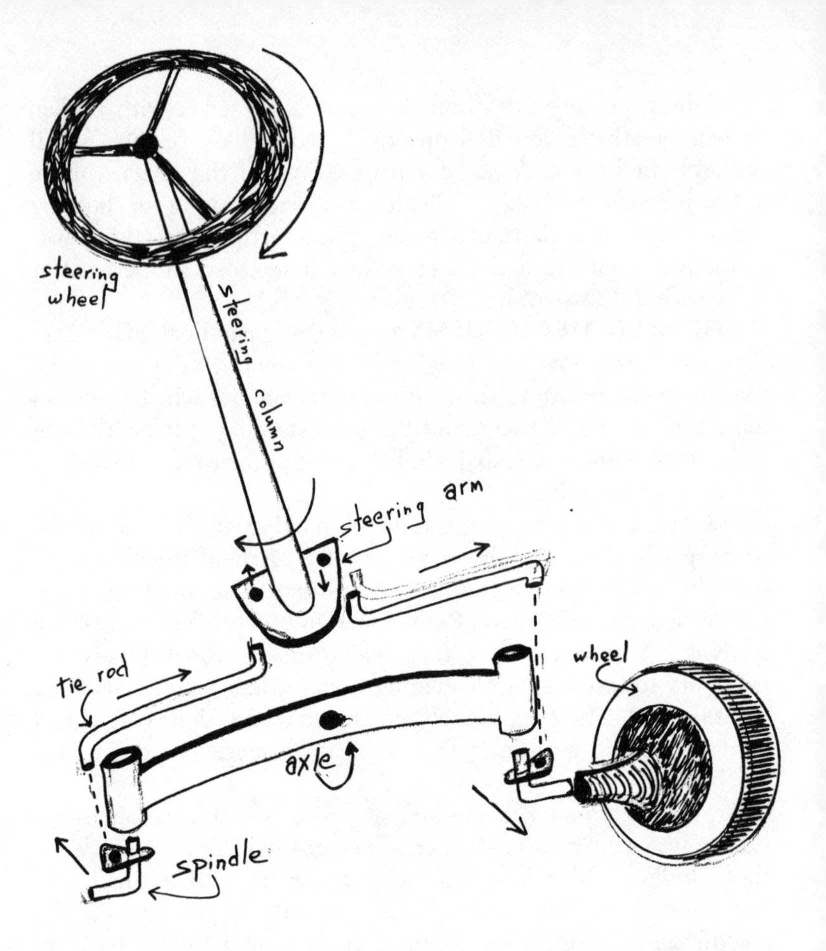

Illustration 8/8. Schematic diagram of typical steering mechanism in Riding Mower.

well as a Ross Steering Gear or a rack-and-pinion if you're lucky. The Ross gear looks like a giant corkscrew; the rack-and-pinion resembles the focussing mechanism on a microscope.

Tie rods can be adjustable rather than simply one piece. Also instead of simply slipping into the steering arm and spindle, there can be ball joints on the tie rods which insure smoother steering on more sophisticated mechanisms.

HARD STEERING is a familiar complaint, especially with the crude steering mechanisms. First check the tires to make sure they are inflated to the recommended pressure. Two opposing needs are involved here. A 600 pound riding lawn mower could ruin a good lawn if it had narrow tires which dig in. Broad tires are needed, therefore, and that often means low pressure inside of them. On the other hand, low-budget steering systems can be very hard to twist into a hard right or hard left with broad, low-pressure tires.

The second easiest place to check as a possible cause of hard steering is the spindle bearings—if bearings they be. If your frequent shot of oil hasn't seeped down, the dry and maybe rusty spindle might be rubbing against an equally dry and rusty surface inside of the axle. In that event you'll have to remove the spindle from the axle, wipe away the dirt and corrosion with emory cloth, and then liberally apply grease. You should renew the grease whenever you perform the periodic maintenance outlined in chapter 2. Better spindles and axles have grease fittings, a feature you might like to add to your fitting-less machine by an afternoon of simple mechanical exercises.

Running into trees, stones and other hard-to-move objects tends to bend, if not break, essential parts of your steering gear such as the spindle arm, the rods, ball joints (if any) and related hardware. Any of these parts, if bent or broken, can cause hard steering, too.

Another occasional cause of hard steering comes from the mistaken notion that front wheels should point straight toward the front. For a variety of good but complicated reasons, the front wheels each should be pointed very slightly inward, a condition called *toe-in*. Measure the distance between wheels at the very front and very back of the front wheels, as illustrated in drawing number 8/9. The front distance should be close to 1/4″ less than the rear distance. John Deere considers 3/16″, toe-in ideal for its small riders. If the tie-rods are adjustable on your

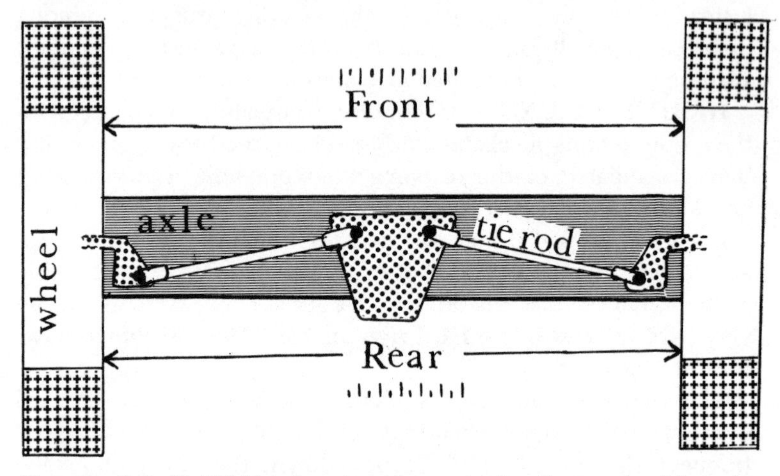

Illustration 8/9. (1) *TOE-IN* is important to easy steering. Measure the distance between the front of the two steering wheels and the distance between the rear. (2) The front should be ¼″ closer together than the rear.

machine, adjust them evenly until the necessary toe-in is obtained.

LOOSE STEERING is generally a sign of misadjustment or wear in one or more of the moving parts already described above in the section on hard steering. The condition can also come about if some essential part has cracked, broken, fallen off or is about to fall off. If the steering *suddenly* feels loose, better check it out promptly.

APPENDIX I

Guide to Lawn Mower Brand Names

THERE are close to 500 different *names for lawn mowers* on sale now or in the recent past. The actual number of *companies selling* lawn mowers is but a fraction of that 500. And the number who *manufacture* lawn mowers or lawn mower parts is smaller still.

This guide can assist you in tracking down the company behind the brand name. In some cases, it is impossible to identify the actual lawn mower assembler, especially in the case of brand names used by department store chains. Department stores often buy equipment from several suppliers and use a single brand name for all of them. In such cases, the name of the chain is given next to the brand name.

Lawn mower assemblers gave very little cooperation in preparing this list. Consequently, directories, both financial and sales, were used to compile a list and to cross-check entries. Every effort has been made to provide accurate and up-to-date information. However, because of frequent and occasionally flagrant inaccuracies found in various trade and financial directories, the author and publisher cannot assume responsibility for errors or omissions which may have crept in.

The left hand column gives the brand name in small-case type or the MANUFACTURER IN CAPITAL LETTERS. In the case of a brand name listing, the right hand column supplies the manufacturer or department store chain using the given brand name. Or, if a MANUFACTURER is listed in the left column, the right gives his address.

EXAMPLE: You own a mower labled "Arlawn" but cannot find the name of its manufacturer. In the left column you will find "Arlawn" listed and next to it is the manufacturer's name, Davis. Looking further, you find a capitalized left-hand column entry "DAVIS, G. W. CORP." and in the right column entry is the address for that manufacturer.

In each instance the address for the brand name or the manufacturer is given only once.

NAME OF MANUFACTURER (Capitalized) or "Brand Name" (in small letters)	Address of a manufacturer or Name of the manufacturer of a "brand name"
"ACI"	American Consolidated Industries Inc.
AMF, INC. (Western Tool Div.)	3811 McDonald Ave. Des Moines, Iowa 50302
"Ace"	Davis, G. W. Corp.
"Aero"	Dille & McGuire
AIR CAP MFG. INC.	P. O. Box 1070 Tupelo, Mis. 38801
AIR-ELECTRIC MACHINE CO., INC.	101 Main St. Lohrville, Iowa 51453
"Air-Flo"	Pro Inc.
"Aladin"	Cooper Mfg. Co.
"All American"	Davis, G. W. Corp.
ALLEGRETTI & CO.	9200 Mason Ave. Chatsworth, Cal. 91311
ALLEN, S. L. & CO. INC.	5th & Glenwood Ave. Philadelphia, Pa.
ALLIS-CHALMERS CORP.	P. O. Box 512 Milwaukee, Wis. 53201
"Ambassador"	AMF, Inc.
AMERICAN CONSOLIDATED INDUSTRIES, INC.	American Consolidated Blvd. Brownsville, Pa. 15417
AMERICAN LAWNMOWER CO.	Muncie, Ind. 47302
"Apache"	Jacobsen Mfg. Co.
"Apollo"	Lambert Inc.
"Argus"	Lambert Inc.
ARIENS CO.	655 W. Ryan St. Brillion, Wis. 54110
"Arlawn"	Davis, G. W. Corp.

"Arlington"	Goodall Div.
"Arrow"	Ariens Co.
"Artisan"	Gamble-Skogmo
"Astro"	American Consolidated Industries Inc.
"Astro-Jet"	Gamble-Skogmo
"Astro-Kut"	Edko Mfg. Inc.
ATACO STEEL PRODUCTS CO.	1040 9th Ave. Grafton, Wis. 53024
"Atlas-Aire"	Atlas Tool & Mfg. Co.
ATLAS LAWN EQUIPMENT CO.	9719 Olive Street Rd. St. Louis, Mo. 63132
ATLAS TOOL & MFG. CO.	5151 Natural Bridge Ave. St. Louis, Mo. 63115
"Autocut"	Goodall Div.
"Automower"	Lawn-Boy
"Avalon"	Pennsylvania Products Div.

B

"BMB"	Vandermolen Corp.
BACHTOLD BROS. INC.	Forrest, Ill. 61741
"Bantam"	Falls Products Inc. Jacobsen Mfg. Co. Sears Mfg.
BARKETT, J. M. MFG. CO.	4210 Dorchester Ave. Charleston, S.C.
"Bearcat"	AMF
"Bellaire"	Atlas Tool & Mfg. Inc.
"Belview"	Goodall Div.
"Bermuda"	Goodall Div.
"Big Red"	Toro Mfg. Co.
"Big 60"	Pennington Mfg. Co.
"Big Snapper"	McDonough Power
"Big Yard Bonanza"	Yazoo Mfg. Co. Inc.
"Blair"	AMF
"Blitzer"	Jacobsen Mfg. Co.
"Blue Diamond"	Pioneer Gen-E-Motor
"Blue Grass"	Dille & McGuire
"Bob-Cat"	Wisconsin Marine
BOLENS DIV.	215 S. Park Port Washington, Wis. 53074

"Bonanza"	General Leisure Products Corp.
"Bonus"	O. M. Scott & Sons
BOTAG MFG. INC.	1490 N.W. 65 Ave. Ft. Lauderdale, Fla. 33313
BREADY TRACTOR & IMPLEMENT CO.	33502 Aurora Rd. Solon, Ohio 44139
"Brenner"	Atlas Tool & Mfg.
BRIGGS & STRATTON	P. O. Box 702 Milwaukee, Wis. 53201
"Broadlawn"	Huffman Mfg. Co.
"Broadmoor"	Goodall Div. Simplicity Mfg. Co.
"Brookline"	Pennsylvania Products Div.
"Brookwood"	Goodall Div.
BUNTON CO.	4303 Poplar Level Rd. Louisville, Ky. 40213
BUSH HOG DIV.	P. O. Box 1039 Selma, Ala. 36701

C

"Cadet"	International Harvester
"Capri"	Lazy Boy
CARR, LELAN, EQUIPMENT CO.	P. O. Drawer 786 Mt. Vernon, Ill. 62864
"Catalina"	AMF
CAUDLE MFG. CO.	7545 N.W. 26 Ave. Miami, Fla. 33147
"Century 20"	Hahn-Eclipse Co.
"Certified"	AMF
"Challenger"	Pioneer Gen-E-Motor
"Charger"	Pennsylvania Products Div.
"Chief"	Jacobsen Mfg. Co. Clinton Engines
"Chieftain"	General Leisure Products Corp.
"Chris-Cut"	Air-Electric Machine Co. Inc.
"Christy"	Air-Electric Machine Co. Inc.
"Cicero"	Power Equipment, Inc.
"Citation"	Huffman Mfg. Co.
CIZEK MFG. & DISTRIBUTING CO., INC.	P. O. Box 86 Clutier, Iowa 52217
"Clean Cut"	Gravely Tractors
CLEAR VUE CO., INC.	1201 W. 6th St. Amarillo, Texas

"Clemson"	Pennsylvania Products Div.
CLINTON ENGINES CORP.	Clark & Maple Maquoketa, Iowa 52060
"Clod-Buster"	Root Mfg. Co. Inc.
"Columbia"	MTD Products Inc.
"Comet"	McDonough Power
"Commando"	Davis, G. W. Corp.
CONSOLIDATED INDUSTRIES, INC.	500 Mixville Rd. West Cheshire, Conn.
"Constellation"	Moto-Mower, Inc.
"Continental"	AMF
CONTINENTAL BELTON CO.	P. O. Box 660 Belton, Texas 76513
"Contour G.P.M."	Kin-Co Mfg.
COOPER MFG. CO.	411 S. 1st Ave. Marshalltown, Iowa 50158
"Coronet"	Davis, G. W. Corp.
"Country Clubber"	Yazoo Mfg. Co. Inc.
"Country Squire"	Moto-Mower, Inc.
"Courier"	Rugg, E. T. Co.
"Craftsman"	Sears Roebuck & Co.
"Croydon"	Pennsylvania Products Div.
CRUE CUT MFG. CO.	4813 Raytown Rd. Kansas City, Mo. 64133
"Custom Hustler"	Pennsylvania Products Div.
"Custom Royal"	Pennsylvania Products Div.
"Cutlass"	McDonough Power
"Cyclo-Mo"	Cooper Mfg. Co.
"Cyclo-Vac"	Cooper Mfg. Co.

D

"Danco"	Danuser Machine Works
DANUSER MACHINE WORKS	1500 Industrial Blvd. Claremont, Okla. 74017
"Dart"	AMF
"David Bradley"	Sears Roebuck & Co.
DAVIS, G.W. CORP.	40 Exchange Pl. New York, N.Y. 10005
DEERE & CO. (John Deere)	John Deere Rd. Moline, Ill. 61265
"Defiance"	Sears Roebuck & Co.

DEINES WELDING, INC.	Hwy. 4 Ransom, Kans. 67572
"Diadem"	Vandermolen Corp.
"Diamond"	Dille & McGuire
DILLE & MC GUIRE	Richmond, Ind.
"Dilly"	Bunton Co.
"Diplomat"	AMF
"Dixie"	Southland Mower
DIXON INDUSTRIES INC.	P.O. Box 494 Coffeyville, Kans. 67337
DRAKE DESIGN, INC.	10127 A Adella Ave. South Gate, Cal. 90280
Dunlap	Sears Mfg. Co. Yardman Div.
"Duo-Master"	Hahn-Eclipse Co.
DURAMATIC PRODUCTS & MFG. INC.	4055 Fitch Rd. Toledo, Ohio 43613
"Duzmore"	Bunton Co.
DYNAMARK CORP.	875 N. Michigan Ave. Chicago, Ill. 60611
"Dynamow"	Dynamark Corp.
"Dyna-Mow"	Air-Cap Mfg. Co.

E

"E-Z-Roll"	Sunshine Mower Co.
"E-Z Wheel"	Rugg, E. T. Co.
EATON CORP.	1101 W. Hanover St. Marshall, Mich. 49068
"Economy"	Engineering Products Co.
"Econoride"	Gamble-Skogmo
"Edge-N-Trim"	Cooper Mfg. Co.
"Edge-R-Trim"	Jacobsen Mfg. Co.
EDKO MFG. INC.	2725 2nd Ave. Des Moines, Lowa 50313
"Electra"	General Leisure Products Corp.
"Emperor"	Ariens
ENGINEERING PRODUCTS CO.	1900 E. Ellis St. Waukesha, Wis. 53186
ESKA CO.	2400 Kerper Blvd. Dubuque, Iowa 52001
"Estate"	Jacobsen Mfg. Co.
"Estate Ace"	Yazoo Mfg. Co. Inc.
"Estate Keeper"	Bolens Div.
EXCEL INDUSTRIES, INC.	Box 385 Hesston, Kan. 67062

F

FAIRBURY PRODUCTS, INC.	P. O. Box 117 Anchor, Ill. 61720
"Fairway"	Ariens Jacobsen Mfg. Co.
"Falcon"	AMF
"Falls"	Sycamore Mfg. Co.
FALLS PRODUCTS, INC.	415 Railroad Av. Genoa, Ill. 60135
"Family"	O. M. Scott & Sons
"Farm & Ranch"	Lazy Boy Lawn Mower Co. Inc.
"Favorite"	Beazley Tractor
"Featherweight"	American Lawn Mower Co.
"Fiftytwo"	Goodall Div.
"Firefly"	Davis, G. W. Corp.
"Flea-Not"	Nott Mfg. Co.
"Fleet Wheel"	Rugg, E. T. Co.
"Fleetwood"	Poloron Prod. of Ind. Inc.
"Flex-N-Float"	Ariens
FLO-MOW MOWER MFG. CO.	700 Blue Ridge Ext. Granview, Mo. 64030
"Flying R"	Root Mfg. Co.
"Folbate"	Middleman, Al Co.
"Forco"	King-O-Lawn Inc.
"4-Blade Rotaries"	Jacobsen Mfg. Co.
"Four Way"	Bunton Co.
"Frederick Atkins"	Davis, G. W. Corp.
FREDERICK MFG. CO., INC.	1406 Agnes Kansas City, Mo. 64127
"Fulton"	Atlas Tool Mfg.
"Fury"	General Leisure Products Corp. Vandermolen Corp.

G

GANNTT MFG. CO.	P. O. Box 49 Macon, Ga. 31202
"Garden Mark"	Montgomery Ward
"Garden Pride"	Southland Mower Co. Inc.
"Gard-N-Yard"	Ariens Co.
"Gemini"	Lambert Inc.
GENERAL ELECTRIC CO.	Bldg. 702, Corporation Park Schenectady, N.Y. 12345

GENERAL LEISURE PRODUCTS CORP.	6200 N. 16th St. Omaha, Nebr. 68101
GILBERT MOWER SALES	P. O. Box 338 Atlanta, Ill. 61723
GILSON BROS. CO.	Box 152 Plymouth, Wis. 53073
"Glendale"	Goodall Div.
GOODALL DIV.	1405 Bunton Rd. Louisville, Ky. 40213
"Gran Prix"	Wheeler Mfg. Co.
"Grass Groomer"	Dille & McGuire
"Grasshopper"	Moridge Mfg. Co.
"Grasshound"	General Leisure Products Corp.
"Grass Jet"	Atlas Tool & Mfg.
"Grassmaster"	Toro Co.
GRAVELY	Gravely Ln. Clemmons, N.C. 27012
"Great American"	Pennsylvania Products Div.
GREAT STATE	818 Webster Shelbyville, Ind. 46176
"Greenbrier"	Goodall Div.
"Greensmower"	Jacobsen Mfg. Co.
"Guardian"	Toro Mfg. Co.
GULF MFG. CORP.	P. O. Box 3727 Panama City, Fla. 32401

H

"HDC"	Vandermolen Corp.
HAHN-ECLIPSE CO.	1625 N. Garvin St. Evansville, Ind. 47717
HAL-GAN PRODUCTS INC.	10129 N. Swan Rd. Mequon, Wis. 53092
"Handiman"	Sears Roebuck & Co.
"Handy Andy"	Davis, G. W. Corp.
HANNIBAL MOWER CORP.	821 Lyon St. Hannibal, Mo. 63401
HARDY SUPPLY CO.	Macksville, Kan. 67557
"Hawk"	AMF
HEILMAN ENTERPRISES INC.	200 W. Marion Mishawaka, Ind. 46544
HENDERSON MFG. CO.	57 Ruby St. Fisher, Ill.

"Hi-Boy"	Lazy Boy Lawn Mower Co. Inc.
"Highlander"	Moto-Mower
"Holiday"	Southland Mower Co. Inc.
HOLLOWAY, L.M. MFG. CO.	10283 Corunna Rd. Swartz Creek, Mich. 48473
HOMELITE DIV.	Box 7047 Charlotte, N.C. 28217
"Homko"	AMF
"Hornet"	Davis, G. W. Corp.
"Huffy"	Huffman Mfg. Co.
HUFFMAN MFG. CO.	P. O. Box 1204 Dayton, Ohio 45401
HULL INDUSTRIES	Hull, Iowa 51239
"Hurricane"	National Metal Products
"Husky"	Bolens Div.
"Hustler"	Excel Industries Inc. Pennsylvania Products Div.
"Hydro-Hustler XL"	Excel Industries Inc.
"Hytamatic"	Huffman Mfg. Co.

I–J

"Imperial"	Ariens Davis, G. W. Corp. Yazoo Mfg. Co. Inc.
"Imperial-26"	Jacobsen Mfg. Co.
INDUSTRIAL PRODUCTS CORP.	Rte. 1, Box 258 Lexington, Va. 34450
"Intermediate"	Sears Mfg. Co.
INTERNATIONAL HARVESTER CO.	401 N. Michigan Ave. Chicago, Ill. 60611
"Islander"	Roto-Hoe & Sprayer Co.
JACOBSEN MFG. CO.	1721 Packard Ave. Racine, Wis. 43403
JAMAKA CO.	349 E. Virginia St. McKinney, Texas 74069
"Javelin Mark II"	Jacobsen Mfg. Co.
"Jet"	Ariens
"Jim Dandy"	Engineering Products Co.
"Juno"	Lambert Inc.

K

KATO EQUIPMENT CO.	P. O. Box 4396 Whittier, Cal. 90605
"Keen Kutter"	Dille & McGuire

F. D. KEES MFG. CO.	700 Park Ave.
	Beatrice, Nebr. 68310
KINCO MFG. CO.	170 N. Pacal
	St. Paul, Minn. 55104
"King"	Dille & McGuire
KING O'LAWN INC.	10127 Adella Ave.
	South Gage, Cal. 90280
KISTLER PRODUCTS INC.	9 Hackett Dr.
	Tonawanda, N.Y. 14150
"Kleen Kool"	Whipper-Clipper
"Kleen-Sweep"	Rugg, E. T. Co.
"Klicker"	Atlas Tool & Mfg. Co.
"Klipper"	Cooper Mfg. Co.
"Klipper-Trim"	Cooper Mfg. Co.
"Kurkee"	Southland Mower Co. Inc.
"Kut-King"	Rowco Mfg. Co.
KUT-KWICK CORP.	1927 Newcastle St.
	Brunwwick, Ga. 31520

L

"Lakeside"	Montgomery Ward
LAMBERT INC.	519 Hunter Ave.
	Dayton, Ohio 45404
"Lancer"	General Leisure Products Corp.
"Landlord"	Simplicity Mfg. Co. Inc.
"Lariat"	Roof Mfg. Co.
"Lark"	AMF
"Lawn Ace"	Moto-Mower
"Lawn Bee"	Sarlo Power Mowers Inc.
"Lawn Boss"	White Outdoor Products
LAWN BOY	Galesburg, Ill. 61401
"Lawn Champ"	Sunbeam Outdoor Co.
LAWN CRAFT MFG. CORP.	Hoffman, Ill. 62250
"Lawn Flite"	MTD Products
"Lawn Keeper"	Bolens Div.
"Lawn King"	Yazoo Mfg. Co. Inc.
"Lawn Lark"	Bunton Co.
"Lawn Prince"	Jacobsen Mfg. Co.
"Lawn Ranger"	Wheel Horse Products Inc.
"Lawn Rov'R"	Heilman Enterprises Inc.
"Lawn Scout"	AMF
"Lawn Vac"	Ariens
"Lawndale"	Pennsylvania Products Div.

"Lawnmowbile"	AMF
"Lawn-O-Vac"	General Leisure Products Corp.
LAZY BOY LAWN MOWER CO., INC.	1315 W. 8th St. Kansas City, Mo. 64101
"Levelawn"	Goodall Div.
"Lexington"	Industrial Products Corp.
"Lightway"	American Lawnmower Co.
"Little Mo"	Bunton Co.
"Little Wheel Wonder"	Yazoo Mfg. Co. Inc.
LITTLE WONDER, INC.	1028 Street Rd. Southampton, Pa. 18966
"Loafer"	Lawn Boy
LOCKE MFG. DIV.	1085 Connecticut Ave. Bridgeport, Conn. 06607

M

MTD PRODUCTS INC.	5389 W. 130th St. Cleveland, Ohio 44111
MAGNA AMERICAN CORP.	P. O. Box 90 Raymond, Miss. 39154
"Manor"	Jacobsen Mfg. Co.
"Manorway"	Ariens
"Marauder"	Lawn Boy
"Mariner"	Lambert Inc.
"Mark III"	Jacobsen Mfg. Co.
"Mark III"	Poloron Products of Ind. Inc.
"Mark 26"	Eska Co. Lambert Inc.
"Mark 30"	Eska Co.
"Marksman"	AMF
MASSEY-FERGUSON, INC.	1901 Bell Ave. Des Moines, Iowa 50315
"Master Mower"	Yazoo Mfg. Co. Inc.
"Master Ride"	Air Cap Mfg. Inc.
"Mastercut"	Air Cap Mfg. Inc.
"Maverick"	General Leisure Products Corp.
MAYO, D. R. SEED CO.	4718 Kingston Pike Knoxville, Tenn. 37919
MC DONOUGH POWER EQUIPMENT INC.	Macon Rd. Macon, Ga. 30253
MC LANE TOOL & DIE	7210 E. Rosecrans Blvd. Paramount, Cal. 90723
MEAD SPECIALTIES CO.	4111 N. Knox Ave. Chicago, Ill. 60641

"Mermaid"	Davis, G. W. Corp.
"Meteor"	Pennsylvania Products Div.
MICHAELS MACHINE CO.	709 N. 19th St. Mattoon, Ill. 61938
MIDDLEMAN, AL	220 W. 42nd St. New York, N.Y. 10036
"Midland"	Lawn Boy
"Mini-Brute"	Roper Sales
"Mini-Sabre"	Pennsylvania Products Div.
MITCHELL MFG. CO.	P. O. Box 66A Wood River Jct., R.I. 02894
"Modern Line"	MTD Products
"Monett"	Mitchell Mfg. Co.
MONO MFG. CO.	Hwy. 160 W. Springfield, Mo. 65804
"Monterey"	Moto-Mower
MONTGOMERY WARD	619 W. Chicago Ave. Chicago, Ill. 60607
MORIDGE MFG. INC.	P. O. Box 637 Moundridge, Kans. 67107
"Moto-Boy"	Moto-Mower
"Moto-Cut"	Moto-Mower
MOTO-MOWER	Cambridge, Ohio 43725
MOTT CORP.	Southview & Shawnut St. LaGrange, Ill.
"Mow-Blow"	Sensation Mfg. Co.
"Mow-R"	Continental Belton Co.
"Mow-Rite"	General Leisure Products Corp.
"Mow-Trac"	AMF
"Mowa-Matic"	General Leisure Products Corp.
MOWETT SALES CO., INC.	110 W. Mason Odessa, Mo. 64076
"Moz-All"	Wind-King Mfg. Co.
MURRAY OHIO MFG. CO.	P. O. Box 268 Brentwood, Tenn. 37027
"Mustang"	Bolens Div. General Leisure Products Corp. Mowett Sales Co. Inc. Yard-Man Inc.

N–O

"National"	American Lawnmower Co.
NATIONAL MOWER CO.	700 Raymond St. Paul, Minn. 55114

"New Favorite"	American Lawnmower Co.
"Oakhill"	Goodall Div.
"Orbit Air"	Bolens Div.

P

"Pacer"	Gilson Bros. Co.
"Pal"	Hahn-Eclipse Co.
"Palomino"	Roof Mfg. Co.
"Panzer"	Virginia Metalcrafters
"Paramount"	Allegretti & Co.
"Park 30"	Jacobsen Mfg. Co.
"Parklane"	Huffman Mfg. Co.
"Parkridge"	Goodall Div.
PARKTON CORP.	N. E. Industrial Park Hwy. 156 & Kynkle Dr. Elsworth, Kans. 67439
"Pathfinder"	Root Mfg. Co.
"Peerless"	Tecumseh Products Clinton Engines
"Pemco"	Power Equipment Mfg.
"Penn"	Virginia Metalcrafters
"Penna-Lawn"	Pennsylvania Products Div.
"Pennant"	Pennsylvania Products Div.
PENNINGTON MFG. CO.	36 Industrial Rd. Addison, Ill.
"Pennsylvania"	Virginia Metalcrafters
PENNSYLVANIA PRODUCTS DIV.	P. O. Box 928 Martinsburg, W.Va. 25401
PERRY CO, The	P. O. Box 7187 Waco, Texas 76710
PHILLIPS DISTRIBUTING CO.	Keller Smithfield Rd. Rte. 1, Box 36-A Keller, Texas 76248
"Pincor"	Pioneer Gen-E-Motor
"Pinehurst"	Goodall Div.
PIONEER GEN-E-MOTOR	5845 W. Dickens Ave. Chicago, Ill. 60639
"Planet-Jr."	Allen, S. L. & Co., Inc.
POLORON PRODUCTS INC.	Box 152 Batesville, Miss. 38606
POWER EQUIPMENT CORP.	1150 Broadway New York, N.Y. 10001

POWER EQUIPMENT MFG. CORP.	1902 Tiger Tail Blvd. Tiger Tail Industries Park Dania, Fla. 33004
"Power King"	Engineering Products Co.
"Power Lawn"	Davis, G. W. Corp.
"Powerama"	General Leisure Products Corp.
"Poweride"	Gamble-Skogmo
"PoweRider"	Whiz-Mow Mfg. Co.
"Powermow"	F. D. Kees Mfg. Co.
"Power-O-Matic"	Power Equipment Corp.
"Pow-R-Boy"	Hahn-Eclipse Co.
"Pow-R-Edger"	Jacobsen Mfg. Co.
"Pow-R-Pro"	Hahn-Eclipse Co.
"Premier"	Lazy Boy Lawn Mower Co., Inc.
"Premium"	Engineering Products Co.
PRO/INC	4225 Pro St. Shreveport, La. 71109
"Pro-Master"	Kato Equipment Co.
"Pro/Mow"	Pro/Inc.
"Pro-Power Mower"	Pro/Inc.
"Pro-Vac"	Kato Equipment Co.

Q–R

"Qualcast"	Middleman, Al Co.
"Qwik-Pik"	Rowco Mfg. Co., Inc.
"Ram"	Jacobsen Mfg. Co.
RAM INDUSTRIES, INC.	401 S. 2nd St. Leavenworth, Kan. 66048
"Ranchero"	Huffman Mfg. Co.
"Range Rider"	Root Mfg. Co.
"Ranger Rider"	Roof Mfg. Co.
"Red Rambler"	Gamble-Skogmo
"Red Rider"	Gamble-Skogmo
"Red Tip"	Whiz-Mow Mfg. Co.
"Redwing"	Bunton Co.
"Renovator"	Jacobsen Mfg. Co.
"Republic"	American Lawn Mower Co.
"Ride King"	Swisher Mower & Machine Co.
"Riviera"	AMF Moto-Mower
"Roamer"	Huffman Mfg. Co.

"Rocket VI"	Ariens
"Rocket 88"	Gamble-Skogmo
ROOF MFG. CO.	1011 W. Howard St. Pontiac, Ill. 61764
ROOT MFG. CO.	127 E. 11th St. Baxter Springs, Kans. 66713
ROPER SALES	1905 W. Court Kankakee, Ill. 60901
"Rota-Shear"	Allegretti & Co.
"Roto-Clipper"	Falls Products Inc.
"Roto-Cutter"	Roto Hoe & Sprayer Co.
ROTO HOE & SPRAYER CO.	Rte. 87 Newbury, Ohio 44065
"Roto-Rugg"	Rugg, E. T. Co.
ROWCO MFG. CO., INC.	48 Emerald St. Keene, N.H. 03431
ROXY-BONNER, INC.	2000 Pioneer Rd. Huntington Valley, Pa. 19006
"Royal"	King O'Lawn Inc. Pennsylvania Products Div.
RUGG, E. T. CO.	1 Sisal St. Newark, Ohio 43055
"Rugged Rider"	Rugg, E. T. Co.

S

"SA Special"	American Lawn Mower Co.
"Sabre"	Pennsylvania Products, Div.
"Samson"	General Leisure Products Corp.
"San Fernando"	Davis, G. W. Corp.
SARLO POWER MOWERS, INC.	2315 Anderson Av. Fort Meyers, Fla. 33902
"Scamper"	Allis-Chalmers
"Scorpio"	Lambert Inc.
SCOTT COMPANY	P. O. Box 1097 Hutchinson, Kan. 67501
SCOTT, O. M. & SONS	Marysville, Ohio 43040
"Scout Deluxe"	Moto-Mower
SEARS MFG. CO.	1718 S. Concord Davenport, Iowa 52802
SEARS ROEBUCK & CO.	303 E. Ohio St.

	Chicago, Ill. 60611
SENSATION MFG. CO.	7577 Burlington St. Ralston, Neb. 68127
"Serf"	Simplicity Mfg. Co. Inc.
"769 GLM"	Goodall Div.
"Shark"	Davis, G. W. Corp.
"Silver Streak"	Frederick Mfg. Co., Inc.
SIMPLICITY MFG. CO., INC.	500 N. Spring St. Port Washington, Wis. 53074
"Sit N'Cut"	Air Cap Mfg. Inc.
"Snapper"	McDonough Power Equipment Inc.
"Snappin' Turtle"	McDonough Power Equipment Inc.
SOLO MOTORS, INC.	Box 5030 Newport News, Va. 23606
SOUTHLAND MOWER CO., INC.	Old Montgomery Hwy. Selma, Ala. 36701
"Sovereign"	Simplicity Mfg. Co. Inc.
SPECIALIZED PRODUCTS, INC.	Rte. 48 East Taylorsville, Ill. 62568
SPEEDEX TRACTOR CO.	367 N. Freedom St. Ravenna, Ohio 44266
"Speedy"	American Lawn Mower Co.
"Spinaway"	Hal-Gan Products Inc.
"Sportlawn"	Toro Mfg. Co.
"Sports Rider"	Moto-Mower
"Sportsman"	Toro Mfg. Co.
"Sprint"	Bolens Div. Roper Sales
"Stallion"	Poloron Products of Ind., Inc. Sensation Mfg. Co.
STANLEY TOOL DIV.	666 Myrtle St. New Britain, Conn.
"Star Clipper"	Caudle Mfg. Co.
"Strider"	Allis-Chalmers
"Suburban"	Bolens Div. General Leisure Products Corp.
"Suburbanite"	Rugg, E. T. Co. Yazoo Mfg. Co. Inc.
"Success"	American Lawn Mower Co.

SUFFOLK IRON FOUNDRY	200 W. 42nd St. New York, N.Y. 10036
"Sun Valley"	Moto-Mower
SUNBEAM OUTDOOR CO.	2001 S. York Rd. Oak Brook, Ill. 60521
SUNSHINE MOWER CO.	4320 15th St. Bradenton, Fla. 33505
"Superchief"	Jacobsen Mfg. Co.
"Super Jet"	Ariens
"Sweep Clean"	Root Mfg. Co.
"Swish-err"	Swisher Mower & Machine Co.
SWISHER MOWER & MACHINE CO.	333 E. Gay St. Warrensburg, Mo. 64093
"Swordfish"	Davis, G. W. Corp.
SYCAMORE MFG. CO.	415 Railroad Ave. Genoa, Ill. 06135

T

TECUMSEH PRODUCTS CO.	Grafton, Wis. 53024
"Temco"	King-O-Lawn Inc.
"Thirty Six"	Goodall Div.
"Thunderbird"	AMF
"Tiger"	Allis-Chalmers
"Titan"	Lambert Inc.
"Tom Boy"	Hannibal Mower Corp.
"Top Flite"	MTD
"Top Way"	Hahn-Eclipse Co.
TORO COMPANY, THE	8111 Lyndale Ave. S. Minneapolis, Minn. 55420
"Town-and-Country"	Yazoo Mfg. Co. Inc.
"Trac-Team"	Ariens
"Tracker"	Allis-Chalmers
"Trail Rider"	Sensation Mfg. Co.
"Trim-O"	Jacobsen Mfg. Co.
TRIMMER LAWN MOWER CO.	Victoria Ave. & 67th St. Los Angeles, Cal. 90066
"Trio-Rig"	Pennsylvania Products Div.
"Triplex"	National Mower Co.
"Triplex Greenes"	Locke Mfg. Div.
"Tru-Cut"	McDonough Power Equipment Inc.

TRU-TEST MFG. CO.	400 Cottage Ave. Carpentersville, Ill. 60110
TUBESING CORP.	129 Maple St. Decatur, Ga. 30030
"Turbocone"	Jacobsen Mfg. Co.
"Turbo-Jet"	Gamble-Skogmo
"Turb-O-Matic"	Power Equipment Inc.
"Turbo-Vac"	Jacobsen Mfg. Co.
"Turbo-Vent"	Jacobsen Mfg. Co.
"Turfmaster"	Huffman Mfg. Co.
TURF-O-MATIC, INC.	9 Monmouth Park Pl. Oceanport, N.J. 07757
"Twelve 7602"	Goodall Div.
"Twentyeight"	Goodall Div.
"Twentyone"	Goodall Div.
"Twentyone SP"	Bunton Co.

U–V

"Ultravac"	Davis, G. W. Corp.
"Universal"	Yazoo Mfg. Co. Inc.
"Vac-Master"	Kato Equipment Co.
"Vacumaire"	General Leisure Products Corp.
"Vac-U-Matic"	General Leisure Products Corp.
"Valiant"	AMF
VANDERMOLEN CORP.	119-T Dorsa Ave. Livingston, N.J. 07039
"Vanguard"	Lambert Inc.
"Victor"	Jacobsen Mfg. Co.
VIRGINIA METALCRAFTERS, INC.	Waynesboro, Va. 22980
"Viscount"	General Leisure Products Corp. Jacobsen Mfg. Co.
"Vista"	Southland Mower Co., Inc.
"Vulcan"	Southland Mower Co., Inc.

W

WARREN'S TURF NURSERY	8400 W. 111 St. Palos Park, Ill. 60464
"Warrior"	General Leisure Products Corp.
"Wasp"	Davis, G. W. Corp.

WHEELER MFG. CORP.	P. O. Box 688 Ashtabula, Ohio 44004
WHEEL HORSE PRODUCTS, INC.	515 W. Ireland Rd. South Bend, Ind. 46614
WHIPPER-CLIPPER CO., INC.	Star Rte. Bonham, Texas 75418
"Whirlwind"	Toro Mfg. Co.
"Whisper Cut"	General Leisure Products Corp.
WHITAKER MFG. CO. DIV.	648 W. Washington East Peoria, Ill. 61611
WHITE OUTDOOR PRODUCTS	2625 Butterfield Rd. Oak Brook, Ill. 60521
WHIZ-MOW MFG. CO.	305 Main St. Warsaw, Ill. 62379
"Wildcat"	AMF
WILLARD MFG. CO., THE	501 Sunset Dr. Meadville, Pa. 16403
"Windsor"	Goodall Div.
WIND-KING MFG. CO.	Merrill, Iowa 51038
"Wingfoot"	American Lawn Mower Co.
WISCONSIN MARINE CO.	Box 28 Lake Mills, Wis. 53551
"Wonder-Boy"	Simplicity Mfg. Co. Inc.
"Work Dodger"	Lazy Boy Lawn Mower Co., Inc.
"Workmaster"	Yazoo Mfg. Co. Inc.

X Y Z

YARD-MAN DIV.	P. O. Box 2741 Cleveland, Ohio 44111
YAZOO MFG. CO., INC.	3607 Livingston Rd. Jackson, Miss. 39216
"Yeoman"	Simplicity Mfg. Co. Inc.
"Zephyr"	Hahn-Eclipse Co.

APPENDIX II

Routine Maintenance Checklist

EVERY USE: BEFORE MOWING

1. Check the level of both gasoline and oil.
2. Look for loose parts on the engine.
   a. The air cleaner.
   b. Oil drain plug.
   c. Mounting bolts.
   d. Spark plug wire.
   e. Fuel line connections.
   f. Gas tank mounting brackets.
   g. Throttle, choke springs, levers on carburetor.
3. Look for loose parts on the mower itself.
   a. Controls.
   b. Drive mechanism.
   c. Seat support bolts.
4. Look for leaks since last using the mower.
5. Oil all of the principal moving parts.

EVERY USE: AFTER MOWING

1. Remove all grass, leaves, dirt and other debris from the mower, engine, and moving parts.
2. Wipe your entire mower clean and dry.
3. Fill the gasoline tank.

PERIODIC MAINTENANCE

1. Charge the battery.
2. Remove grass and other debris from under the engine blower housing.
3. Change the oil.
4. Clean the air filter.
5. Check and clean ignition points (unless located beneath flywheel).
6. Clean the spark plug and set it to the proper gap.
7. Clean away carbon from inside the cylinder.
8. Check the pulleys, belts, chains, sprockets and clutches.
9. Oil and grease all moving parts.
10. Sharpen and balance the blade or lap the reel, if necessary.

AT THE END OF EVERY SEASON

1. Drain the gas tank and purge fuel from the engine.
2. Change the oil.
3. Remove the spark plug and add one tablespoon of clean oil through the spark plug hole.
4. Oil all moving parts.
5. Dry and then oil all metal parts.
6. Charge any storage battery and disconnect it to prevent discharge. Clean the battery case.
7. Block up the tires.

AT THE BEGINNING OF EVERY SEASON

1. Take away the tire blocks.
2. Sharpen the rotary blade or lap the reel.
3. Oil and grease all moving parts.
4. Charge the battery.
5. Clean deposits off electrical parts.
6. Drain the oil and replace with fresh oil.
7. Fill the fuel tank with fresh gasoline.
8. Start the engine to let oil burn away.
9. Tune the engine.

APPENDIX III

Safety Suggestions of the American National Standards Institute, Inc.

MOWER PURCHASE

1. Be sure that the mower you buy conforms to the requirements of the American National Standards Institute, Inc.
2. Check to see that you have detailed starting and operating instructions.

TRAINING

1. Never allow children to operate a power mower.
2. Learn how to stop the mower quickly.
3. Instruct children to stay away from the mower while it is in operation.

PREPARATION

1. Before starting, clear area of debris.
2. Set mower at highest cutting height when mowing in rough ground or in tall weeds.
3. Mow only in daylight or in good artificial light.
4. Do not operate mower in wet grass.
5. Do not operate mower with guards removed.
6. Wear substantial shoes and long pants while using mower.
7. Fill gas tank outdoors, but never while engine is running. Avoid spilling
8. Check power cords on electric mowers for cracks or breaks.

OPERATION

1. Give complete and undivided attention to the job at hand.
2. Do not operate mower in the vicinity of other persons.
3. Stop the motor or engine whenever you leave the mower operating position.
4. Operate the engine at the slowest speed that will cut satisfactorily.
5. Stand clear of the front of self-propelled mowers.
6. Use caution when operating the mower on uneven terrain. Maintain good footing.
7. Do not let others ride with you on riding mowers.
8. Check for breakage and repair any damage after striking foreign object.

MAINTENANCE AND STORAGE

1. Follow maintenance instructions given by manufacturer.
2. Have a competent serviceman inspect the mower each year.
3. Disconnect the spark plug wire before making any adjustment or repair.
4. Keep engine or motor free of grass or debris buildup.
5. Store gasoline in an approved metal container in a cool, dry place.
6. A well-maintained mower operated correctly will produce best mowing results.

Index